Lecture Notes in Mathematics

1647

Editors:
A. Dold, Heidelberg
F. Takens, Groningen

Springer
Berlin
Heidelberg
New York
Barcelona
Budapest
Hong Kong
London
Milan
Paris
Santa Clara
Singapore
Tokyo

Danielle Dias   Patrick Le Barz

Configuration Spaces over Hilbert Schemes and Applications

Springer

Authors

Danielle Dias
Patrick Le Barz
Laboratoire de Mathématiques
Université de Nice – Sophia Antipolis
Parc Valrose
F-06108 Nice, France
e-mail: ddias@math.unice.fr
        lebarz@math.unice.fr

Cataloging-in-Publication Data applied for

Die Deutsche Bibliothek - CIP-Einheitsaufnahme

Dias, Danielle:
Configuration spaces over Hilbert schemes and applications /
Danielle Dias ; Patrick LeBarz. - Berlin ; Heidelberg ; New
York ; Barcelona ; Budapest ; Hong Kong ; London ; Milan ;
Paris ; Santa Clara ; Singapore ; Tokyo : Springer, 1996
  (Lecture notes in mathematics ; 1647)
  ISBN 3-540-62050-8
NE: LeBarz, Patrick:; GT

Mathematics Subject Classification (1991): 14C05, 14C17

ISSN 0075-8434
ISBN 3-540-62050-8 Springer-Verlag Berlin Heidelberg New York

Typesetting: Camera-ready T_EX output by the authors
SPIN: 10520222   46/3142-543210 - Printed on acid-free paper

Table of Contents

Introduction

0.1

Let $f : V \longrightarrow W$ be a morphism of non-singular varieties over $\mathbb{C}$, with $\dim V < \dim W$. Let $d = \mathrm{cod}(f) = \dim W - \dim V$. The locus V_k of elements $x \in V$ such that there exists at least $(k-1)$ other points of V in the fiber $f^{-1}f(x)$ is called k-*uple locus of* f. When it exists, a class m_k in the Chow ring $\mathrm{CH}^{\bullet}(V)$ of V, which represents V_k, is called k-*uple class of* f. Then a k-*uple formula* is a polynomial expression which gives m_k in terms of the Chern classes c_i of the virtual normal bundle $\nu(f) = f^*TW - TV$.

0.2

When one deals with the double formula, one is interested in the set of elements $x \in V$ such that the fiber $f^{-1}f(x)$ contains at least one other point in addition to x. A typical example is the imbedding with normal crossings $f : C \longrightarrow \mathbb{P}^2$ of a smooth curve, where one wishes to count the number of double points of $f(C)$.

The case $k = 2$ was treated thoroughly by Laksov ([La]). The double formula was also found by Ronga ([Ro]) in the C^∞ case. The demonstration consists in looking at the blowing-up $\widetilde{V \times V}$ of $V \times V$ along the diagonal and applying the residual intersection formula ([FU2], thm 9.2, pp 161–162) in order to remove the exceptional divisor which corresponds to the solutions $x_1 = x_2$ of $f(x_1) = f(x_2)$ (which we do not want) at the lifted double locus (see [FU2], pp 165–166). Then the double class m_2 is given in the Chow ring of V by the *double formula* :

$$m_2 = f^*f_*[V] - c_d \, ,$$

where c_d is the dth Chern class of $\nu(f)$, defined above.

0.3

When one deals with the triple formula, one is interested in the set of elements $x \in V$ such that $f^{-1}f(x)$ contains at least two points in addition to x. A typical example is the imbedding with normal crossings $f : S \longrightarrow \mathbb{P}^2$ of a non-singular surface, where one wishes to count the number of triple points of $f(S)$.

One difficulty is to <u>define</u> a class $m_3 \in \mathrm{CH}^{\bullet}(V)$ representing the set V_3 and to compute this class so that one has the *triple formula* :

$$m_3 = f^* f_* m_2 - 2c_d m_2 + \sum_{j=1}^{d} 2^j c_{d-j} c_{d+j} \; ,$$

where c_i is the ith Chern class of $\nu(f)$.

This was done by Kleiman [KL1, KL2], modulo some general hypotheses on the morphism f. Kleiman even established a stronger formula [KL3]. So did Ronga [Ro] in the $\mathcal{C}^\infty$ case. (These are "refined" formulas in the sense that if $f(x_1) = f(x_2) = f(x_3)$, they count the set of non ordered $\{x_1, x_2, x_3\}$ having the same image by f and not the set of ordered (x_1, x_2, x_3) ; therefore there is a gain of 3! in the formulas. The present work, despite the use of Hilbert schemes, will only deal with non refined formulas.

However, all the demonstrations of triple formulas use general hypotheses on the morphism f, essentially the regularity of some "derivative" applications . See also the paper by Colley [Co].

0.4

The goal of the first part of this book is to establish the triple formula without any hypotheses on the genericity of f. Of course, one must immediately :

(i) make it clear that this requires to choose an ad hoc definition of m_3,

(ii) emphasize that in the degenerate case where the triple locus is too big, the formula does not mean much !

Looking for the triple locus of f means looking for the set of (x_1, x_2, x_3) of $V \times V \times V$ such that $f(x_1) = f(x_2) = f(x_3)$. Once again, one wishes to eliminate the solutions with $x_1 = x_2$ or $x_2 = x_3$ or $x_3 = x_1$. One must find a "good" space of triples for V : a space where the locus to be eliminated is a Cartier's divisor. In [KL1], Kleiman uses the space $'Hilb^2(V) \times_V {}'Hilb^2(V)$ where $'Hilb^2(V)$ denotes the universal two-sheeted cover of $Hilb^2(V)$. In [Ro], Ronga blows-up in $Hilb^2(V) \times V$ the tautological $'Hilb^2(V)$. Our suggestion here is to use the space $\widehat{H^3(V)}$ of completely ordered triples of V, introduced in [LB1], which is birational to $V \times V \times V$. Let us recall briefly the construction of $\widehat{H^3(V)}$:

An element $\hat{t} = (p_1, p_2, p_3, d_{12}, d_{23}, d_{31}, t)$ in the product $V^3 \times [Hilb^2(V)]^3 \times Hilb^3(V)$ is a complete triple if it verifies the relations :

$$\begin{cases} p_i \subset d_{ij} \subset t & (\text{ scheme-theoretic inclusions }) \\ p_j = Res(p_i, d_{ij}) \\ p_k = Res(d_{ij}, t) & \text{with} \quad \{i, j, k\} = \{1, 2, 3\} \end{cases}$$

where $Res(\eta, \xi)$ denotes the residual closed point of the $(k-1)$-uplet η contained in the k-uplet ξ.

The motivation is that this space appears to be more natural, in view of the action of the symmetric group S_3. However, one must realize that one ends up computing in $'Hilb^2(V) \times V$ in the process of the demonstration. In particular, the origin of the 2^j that one finds in the triple formula stems from the computation (see § 3.2.4) of the virtual normal bundle of the morphism $'Hilb^2(V) \to Hilb^2(V)$, that one already finds in [KL1] and [Ro].

0.5

Once the space of triples we work with has been chosen, we work along the same lines as Ran [Ra] and Gaffney [Ga] :

(i) if $X \subset Z$ is a non-singular subvariety of a non-singular variety Z, one gives the fundamental class $[\widehat{H^3(X)}]$ in the Chow ring $\mathrm{CH}^\bullet(\widehat{H^3(Z)})$ (theorem 3).

(ii) if $f : V \to W$ is a morphism, one defines the triple class $m_3 \in \mathrm{CH}^\bullet(V)$ as the direct image of the cycle

$$M_3 = [\widehat{H^3(\Gamma_f)}] \cdot [\widehat{H^3(V)} \times W] \quad \in \quad \mathrm{CH}^\bullet(H^3(\widehat{V \times W}))$$

where Γ_f is the graph of f. Tedious but straightforward computations lead to the triple formula (theorem 4).

However, one must realize that in the case where the morphism f has S_2-singularities, the scheme-theoretic intersection $\widehat{H^3(\Gamma_f)} \cap (\widehat{H^3(V)} \times W) \subset H^3(\widehat{V \times W})$ has automatically excess components. This makes the interpretation of the formula tricky. In enumerative geometry, one often gives a formula which is valid in general, even if one must explain afterwards how many "improper" solutions must be removed in order to find the number of "proper" solutions. We have chosen to follow this approach, i.e. we provide a formula which is valid in general, but we are aware that the second half of our task would be to interpret this formula in the degenerate cases, which one cannot avoid. This is done in a preliminary way in chapter 5, where one considers the simplest case where $f : V \to W$ is a morphism from a surface to a volume with S_2-singularity.

0.6

The second part of this book is devoted to the construction of a variety of complete quadruples in order to define a class m_4 in the Chow ring $\mathrm{CH}^\bullet(V)$. The goal is to construct a "good" space of completely ordered quadruples of V, in which the locus to be eliminated is a Cartier's divisor. To do so, one wishes to generalize the construction of the variety $\widehat{H^3(V)}$ of the complete triples. Therefore, the question is to construct *naturally* a variety $\widehat{H^4(V)}$ consisting of ordered quadruples of V, possessing a birational morphism :

$$\widehat{H^4(V)} \rightarrow V \times V \times V \times V \quad,$$

compatible with the action of the symmetric group S_4, and an order-forgetting morphism :

$$\widehat{H^4(V)} \rightarrow Hilb^4(V)$$

The construction of this variety must also be compatible with closed imbeddings : if $V \subset W$ is a subvariety of W, then $\widehat{H^4(V)}$ can be identified with a subvariety of $\widehat{H^4(W)}$.

0.7

A *naïve* generalization $\widehat{H^4_{na\ddot{v}e}(V)}$ of the construction of $\widehat{H^3(V)}$ is not sufficient, as was already pointed out by Fulton ([FU1]). The variety $\widehat{H^4_{na\ddot{v}e}(V)}$ is defined as a subvariety of the product $V^4 \times [Hilb^2(V)]^6 \times [Hilb^3(V)]^4 \times Hilb^4(V)$. To do this, one introduces the following notation :

Notation 1 : If ξ is a point in $Hilb^d(V)$, one will denote by $\mathcal{I}_\xi$ the ideal sheaf of $\mathcal{O}_V$ which defines the corresponding subscheme.

An element $(p_1, p_2, p_3, p_4, d_{12}, d_{13}, d_{14}, d_{23}, d_{24}, d_{34}, t_1, t_2, t_3, t_4, q)$ in the above product is a complete *naïve* quadruple if it verifies the relationships :

$$\begin{cases} \mathcal{I}_{p_i} \cdot \mathcal{I}_{p_j} \subset \mathcal{I}_{d_{ij}} \subset \mathcal{I}_{p_i} \cap \mathcal{I}_{p_j} & \\ \mathcal{I}_{p_k} \cdot \mathcal{I}_{d_{ij}} \subset \mathcal{I}_{t_l} \subset \mathcal{I}_{p_k} \cap \mathcal{I}_{d_{ij}} & \text{for } \{i,j,k,l\} = \{1,2,3,4\}, \\ \mathcal{I}_{p_i} \cdot \mathcal{I}_{t_i} \subset \mathcal{I}_q \subset \mathcal{I}_{p_i} \cap \mathcal{I}_{t_i} & i < j \text{ and } k < l \\ (*) \quad \mathcal{I}_{d_{ij}} \cdot \mathcal{I}_{d_{kl}} \subset \mathcal{I}_q \subset \mathcal{I}_{d_{ij}} \cap \mathcal{I}_{d_{kl}} & \end{cases}$$

Unfortunately, $\widehat{H^4_{na\ddot{v}e}(V)}$ is reducible and singular. We will see below that the conditions $(*)$ introduce excess components.

Recall [I2, F] that (for $\dim V = 3$) the Hilbert scheme $Hilb^4(V)$ is irreducible and singular at the quadruplets q of ideal $\mathcal{M}_p^2$ where $\mathcal{M}_p$ is the ideal of a closed point p of V. Consider the subvariety $R(V)$ of $Hilb^2(V) \times Hilb^4(V) \times Hilb^2(V)$ consisting of elements (d, q, d') satisfying the relations :

$$\mathcal{I}_d \cdot \mathcal{I}_{d'} \subset \mathcal{I}_q \subset \mathcal{I}_d \cap \mathcal{I}_{d'}$$

The variety $R(V)$ possesses a projection onto $Hilb^4(V)$, which we denote by Π. We will see (chapter 6) that $R(V)$ is reducible : $R(V)$ is the union of two irreducible components, one of which dominates $Hilb^4(V)$ by Π. When $\dim V = 3$, for example, we will see that the non dominant excess component is smooth of dimension 8. We will give a geometric description of this extra component (§ 6.4.2, p. 100). The other component, of dimension 12 ($= \dim\ Hilb^4(V)$), is the closure of the graph of the *residual* rational application *Res*, which is defined below. Let us first introduce the following definition :

Definition 1 : One defines the incidence variety $I(V)$ as subvariety of the product $Hilb^2(V) \times Hilb^4(V)$ by the condition :

$$(d, q) \in I(V) \text{ if and only if } d \text{ is subscheme of } q.$$

The second projection
$$\Pi_2 : \quad I(V) \quad \to \quad Hilb^4(V)$$
$$(d, q) \quad \mapsto \quad q$$
is generically a $6 = \begin{pmatrix} 4 \\ 2 \end{pmatrix}$-sheeted cover.

Definition 2 : A triplet $t \subset V$ is said to be amorphous (cf. [KL4]) if its support is reduced to only one point p and if the ideal of t is the square of the ideal of p in a germ of a smooth surface containing p.

In this variety $I(V)$, the *complementary doublet* of d in q is not always well defined. Indeed, let W be the locally closed subvariety of $I(V)$ consisting of elements (d, q) such that :

$- q$ is the union of an amorphous triplet t of support p and a simple point m distinct from p,

$- d$ is the simple doublet $p \cup m$.

At such elements $(d, q) = (p \cup m, t \cup m)$ of $I(V)$, one cannot define the complementary doublet d' because the closed point p does not define canonically a doublet d' in the amorphous triplet t (cf. [ELB]). However, outside of $\overline{W}$, the closure of W in $I(V)$, one can always define the complementary doublet d' of d in q. One gives the following definition of the *residual* rational application :

Definition 3 : Let *Res* be the residual rational application :
$$Res: \quad I(V) \quad \cdots \to \quad Hilb^2(V)$$
$$(d, q) \quad \cdots \to \quad d' = q \setminus d$$
where d' denotes the "other" doublet, once the doublet d in q has been fixed. The ideal which defines the subscheme d' of V is given by the ideal $\mathrm{Ann}(\mathcal{I}_d/\mathcal{I}_q)$ of $\mathcal{O}_V$. Then this ideal verifies the inclusions :

$$\mathcal{I}_d \cdot \mathcal{I}_{d'} \subset \mathcal{I}_q \subset \mathcal{I}_{d'} \cap \mathcal{I}_d \quad .$$

Let U denote the open subset of $I(V)$ on which the *residual* rational application *Res* is well defined. The open set U contains the dense open subset (cf. proposition 3) consisting of elements (d, q) such that the quadruplet q is a simple quadruplet.

Definition 4 : If Res_U denotes the restriction of the rational application *Res* to U, the graph of the regular application $Res_U : U \to Hilb^2(V)$ is called the graph of the rational application *Res*.

Then, the inclusion

$$\Gamma_{Res}(= \Gamma_{Res_U}) \subset U \times Hilb^2(V)$$

holds.

Then the subvariety $R(V)$ of $Hilb^2(V) \times Hilb^4(V) \times Hilb^2(V)$, whose elements (d, q, d') satisfy the inclusions

$$\mathcal{I}_d \cdot \mathcal{I}_{d'} \subset \mathcal{I}_q \subset \mathcal{I}_d \cap \mathcal{I}_{d'}\quad,$$

contains Γ_{Res}, the graph of the application *Res* :

$$\Gamma_{Res} \subset R(V) \subset Hilb^2(V) \times Hilb^4(V) \times Hilb^2(V)$$

Therefore, this subvariety $R(V)$ contains $\overline{\Gamma}_{Res}$, the closure of Γ_{Res} in the product $Hilb^2(V) \times Hilb^4(V) \times Hilb^2(V)$. The following notation is then introduced :

Notation 2 : Let $B(V) = \overline{\Gamma}_{Res}$ be the closure of the graph of the rational application *Res*. One has the inclusions $B(V) \subset R(V) \subset Hilb^2(V) \times Hilb^4(V) \times Hilb^2(V)$. The variety $R(V)$ possesses a projection Π onto $Hilb^4(V)$. Its restriction to $B(V)$ will be denoted by π. If $q \in Hilb^4(V)$ is a quadruplet, $\tilde{q}$ will denote an element of the fiber $\pi^{-1}(q) \subset B(V)$:

$$
\begin{array}{ccc}
\tilde{q} & B(V) & \subset R(V) \\
\downarrow & \downarrow \pi & \swarrow \Pi \\
q & Hilb^4(V) &
\end{array}
$$

It will be shown in chapter 6 that the variety $B(V)$ is irreducible, smooth, of dimension $4 \cdot \dim V$. The tool for the proof consists in using local coordinates for $Hilb^2(V)$ and $Hilb^4(V)$ in order to give local equations for $B(V)$.

0.8

Let us see how a construction of the variety of complete quadruples as subvariety of the product $[\widehat{H^3(V)}]^4 \times [B(V)]^6$ appeared to be natural to us. Consider the simple quadruplet $q \in Hilb^4(V)$, which is the union of four distinct points p_1, p_2, p_3, p_4 of V. The element $\hat{q}$ in $\widehat{H^4(V)}$ constructed from the point (p_1, p_2, p_3, p_4) of V^4 consists in the data of :

   - the four complete triples $\hat{t}_1, \hat{t}_2, \hat{t}_3, \hat{t}_4$ of $\widehat{H^3(V)}$, constructed from the triplets t_1, t_2, t_3, t_4 contained in the quadruplet q. (The notation t_i means that the triplet t_i is disjoint from the simple point p_i.) The complete triple $\hat{t}_i$ corresponds to the point $(p_1, \cdots, \check{p}_i, \cdots, p_4)$ of V^3 ;

   - the six elements $(d_{ij}, q, d'_{ij})_{\{i,j\} \subset \{1,2,3,4\}}$ of $B(V)$ constructed from q, where d_{ij} denotes the simple doublet which is the union of the two points p_i, p_j, and d'_{ij} is the *residual* doublet of d_{ij} in the quadruplet q.

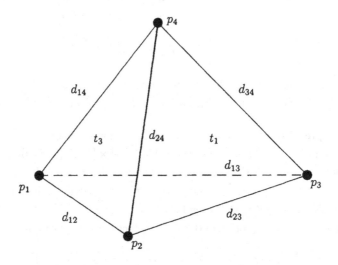

An element $\hat{q}$ of $\widehat{H^4(V)}$ is defined as an element $(\hat{t}_1, \hat{t}_2, \hat{t}_3, \hat{t}_4, \tilde{q}_{12}, \tilde{q}_{13}, \tilde{q}_{14}, \tilde{q}_{23}, \tilde{q}_{24}, \tilde{q}_{34}) \in [\widehat{H^3(V)}]^4 \times [B(V)]^6$, where

$$
\begin{aligned}
\hat{t}_1 &= (\mathbf{p_2, p_3, p_4}, \mathbf{d_{23}, d_{34}, d_{24}}, t_1) \\
\hat{t}_2 &= (P_1, P_3, P_4, D_{13}, D_{34}, D_{14}, t_2) \\
\hat{t}_3 &= (\mathcal{P}_1, \mathcal{P}_2, \mathcal{P}_4, \mathcal{D}_{12}, \mathcal{D}_{24}, \mathcal{D}_{14}, t_3) \\
\hat{t}_4 &= (p_1, p_2, p_3, d_{12}, d_{23}, d_{13}, t_4) \\
\tilde{q}_{ij} &= (\delta_{ij}, q_{ij}, \delta'_{ij}) \quad \text{for } \{i, j\} \subset \{1, 2, 3, 4\}
\end{aligned}
$$

are such that :

1. All the quadruplets q_{ij} are equal to a same quadruplet q,

2. The doublets verify the equalities $\delta'_{ij} = \delta_{kl}$, for $\{i, j, k, l\} = \{1, 2, 3, 4\}$

3. The points verify the equalities :

$$
\begin{aligned}
p_1 &= P_1 = \mathcal{P}_1 \\
p_2 &= \mathcal{P}_2 = \mathbf{p}_2 \\
p_3 &= P_3 = \mathbf{p}_3 \\
\mathcal{P}_4 &= P_4 = \mathbf{p}_4
\end{aligned}
$$

4. The doublets verify the conditions :

$$
\begin{aligned}
d_{12} &= \mathcal{D}_{12} = \delta_{12} \\
d_{23} &= \mathbf{d}_{23} = \delta_{23} \\
d_{13} &= D_{13} = \delta_{13} \\
\mathcal{D}_{24} &= \mathbf{d}_{24} = \delta_{24} \\
\mathcal{D}_{14} &= D_{14} = \delta_{24} \\
D_{34} &= \mathbf{d}_{34} = \delta_{34}
\end{aligned}
$$

5. The triplets verify the scheme-theoretic inclusions $t_i \subset q$. For $i = 1, 2, 3$, the points satisfy the conditions : $p_i = Res(t_i, q)$ and $\mathbf{p}_4 = Res(t_4, q)$.

Chapter 7 will be devoted to the construction of this variety. As before, the construction of this subvariety $\widehat{H^4(V)} \subset [\widehat{H^3(V)}]^4 \times [B(V)]^6$ will be completely explicit. We will see that such a construction enables us to remove the excess components of $\widehat{H^4_{na\ddot{i}ve}(V)}$ (defined in 0.7). The variety $\widehat{H^4(V)}$ constructed in this way is irreducible but unfortunately it is still singular.

A geometric description of the singular locus of $\widehat{H^4(V)}$ will be given in section 7.2.1.

Part one

Double and triple points formula

Chapter 1

Conventions and notation

1.1 Fundamental facts

If X is a non-singular $\mathbb{C}$-variety, its Chow ring graded by the codimension is denoted by $\mathrm{CH}^\bullet(X)$. If $\alpha \in \mathrm{CH}^\bullet(X)$, $\{\alpha\}^k$ denotes the part of <u>codimension</u> k. Of course, these two rules apply :

$(\mathcal{R}_1)$ If x is of pure codimension r, then $\quad x \cdot \{\alpha\}^k = \{x \cdot \alpha\}^{k+r} \quad$.

$(\mathcal{R}_2)$ If $f : X \to X'$ is a proper morphism and if $\mathrm{cod}(f)$ denotes $\dim X' - \dim X$, then :

$$f_*\{\alpha\}^k = \{f_*\alpha\}^{k+\mathrm{cod}(f)}$$

The part of <u>dimension</u> k of $\alpha \in \mathrm{CH}_\bullet(X)$, graded by the dimension, will be also denoted by $\{\alpha\}_k$. Note that :

$(\mathcal{R}_3)$ If f is a proper morphism, $\quad f_*\{\alpha\}_k = \{f_*\alpha\}_k \quad$.

1.2 Conventions

From [FU2], p. 13, if $\Phi : Y \hookrightarrow Y'$ is the canonical imbedding of a subvariety and $\alpha \in \mathrm{CH}^\bullet(Y)$, one will often write α in $\mathrm{CH}^\bullet(Y')$, instead of $\Phi_*\alpha$, which would be more correct.

The fundamental class of a variety X is denoted by $[X]$ or sometimes by $\mathbf{1}$. Improperly but for the sake of simplicity, from § 3.2.4 on, if $Y \subset Y'$ is a subvariety, one will denote by Y and not by $[Y]$, the class of the associated cycle in $\mathrm{CH}^\bullet(Y')$.

Finally, one writes "from (FP)" each time one uses the projection formula :

$$f_*(\alpha \cdot f^*\beta) = f_*\alpha \cdot \beta \tag{FP}$$

1.3   Notation

One denotes by $\theta : \widehat{H^2(X)} \to H^2(X)$ the universal two-sheeted cover of $H^2(X) = Hilb^2(X)$. The scheme $\widehat{H^2(X)}$ is non-singular and possesses two canonical morphisms, which are submersions :

$$\pi_1, \pi_2 : \widehat{H^2(X)} \longrightarrow X .$$

An element $\hat{d}$ of $\widehat{H^2(X)}$ can be identified with a couple (p_1, d) consisting of a point and a doublet containing this point. One can denote by $p_2 = Res(p_1, d)$ the "other point" (see [LB1]). One has a natural isomorphism :

$$\widehat{H^2(X)} \longrightarrow \widetilde{X \times X}$$

with the blowing-up $X \times X$ along the diagonal, which commutes with the projections onto X. This is why one will also write $\hat{d} = (p_1, p_2, d)$.

In $\widehat{H^2(X)}$, there is the "exceptional" divisor F, consisting of the couple (p_1, d) with supp(d) reduced to only one point ; it corresponds to the exceptional divisor of $\widetilde{X \times X}$ via the above mentioned isomorphism. Finally, one checks that :

F is the locus of ramification of the cover $\theta : \widehat{H^2(X)} \longrightarrow H^2(X)$.   (1.1)

Chapter 2

Double formula

2.1   The class of $\widehat{H^2(X)}$ in $\widehat{H^2(Z)}$

Let Z be a non-singular variety of dimension z and $j : X \hookrightarrow Z$ the canonical imbedding of a subvariety of dimension x. One can identify in a natural way the variety $\widehat{H^2(X)}$ of dimension $2x$ with a subvariety of $\widehat{H^2(Z)}$, which is of dimension $2z$. Our purpose is to compute the fundamental class $[\widehat{H^2(X)}]$ in $\mathrm{CH}^\bullet(\widehat{H^2(Z)})$.

Let Π_1 and $\Pi_2 : \widehat{H^2(Z)} \to Z$ denote the two canonical morphisms. In $I = \Pi_1^{-1}(X) \cap \Pi_2^{-1}(X)$ one has the excess component :

$$\mathcal{D} = F \cap \Pi_1^{-1}(X) = F \cap \Pi_2^{-1}(X)$$

where $F \subset \widehat{H^2(Z)}$ denotes, as said already, the exceptional divisor of $\widehat{H^2(Z)}$. The component $\mathcal{D}$ consists of the (d, p_1) with $supp(d) = p_1 \in X$.

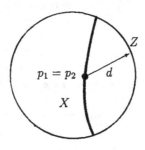

Let us consider the commutative diagram where the restriction $\Pi_{i|\Pi_i^{-1}(X)}$ is denoted by Π_i' and the other arrows are the canonical imbeddings (the parentheses indicate the dimensions) :

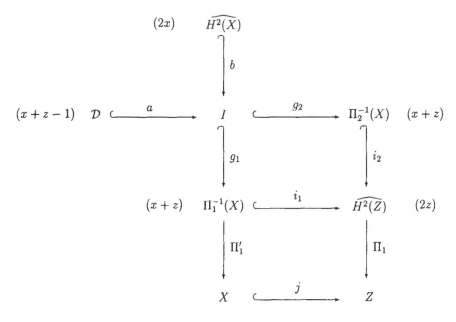

Diagram 1

From [FU2], theorem 9.2, p. 161 and corollary 9.2.2, p. 163, one has in $CH_{2x}(I)$:

$$[\widehat{H^2(X)}] = [\Pi_1^{-1}(X)] \cdot [\Pi_2^{-1}(X)] - \{cN \cdot a_* s(\mathcal{D}, \Pi_2^{-1}(X))\}_{2x} \ ,$$

where $N = g_1^* \nu(\Pi_1^{-1}(X), \widehat{H^2(Z)})$ is the restriction of the normal bundle to $\Pi_1^{-1}(X)$ in $\widehat{H^2(Z)}$.

Hypothesis 1 : From now on, one assumes that the total Chern class $c\nu(X, Z)$ of the normal bundle at X in Z can be written as $j^* c'$ where $c' \in CH^\bullet(Z)$.

Since Π_1 and Π_2 are flat morphisms (see (4.5) and (4.6) in § 4.2), one has :

$$[\Pi_i^{-1}(X)] = \Pi_i^*[X] \qquad\qquad i = 1,\, 2 \qquad\qquad (2.1)$$

Moreover, the normal bundle $\nu(\Pi_1^{-1}(X), \widehat{H^2(Z)})$ can by identified with $\Pi_1'^* \nu(X, Z)$. Hence the total Chern class N is :

$$cN = g_1^* c\nu(\Pi_1^{-1}(X), \widehat{H^2(Z)}) = g_1^* \Pi_1'^* c\nu(X, Z) = g_1^* \Pi_1'^* j^* c' = g_2^* i_2^* \Pi_1^* c' \ ,$$

considering the commutativity of diagram 1. We also have the following lemma :

Lemma 1 *The intersection $\mathcal{D} = F \cap \Pi_2^{-1}(X)$ is transverse.*

Proof :

   See (4.20) in section 4.6.   □

As a result, one gets the equality between Segre classes (inverse of Chern classes) :

$$g_{2*}a_*s(\mathcal{D}, \Pi_2^{-1}(X)) = i_2^*s(F, \widehat{H^2}(Z)) \quad .$$

Notation 3 : One writes $s(F)$ for $s(F, \widehat{H^2}(Z))$ and $s(\mathcal{D})$ for $s(\mathcal{D}, \Pi_2^{-1}(X))$.

Considering the above results, one can rewrite in $CH_{2x}(\Pi_2^{-1}(X))$:

$$g_{2*}\{cN \cdot a_*s(\mathcal{D})\}_{2x} = g_{2*}\{g_2^* i_2^* \Pi_1^* c' \cdot a_*s(\mathcal{D})\}_{2x}$$
$$= \{i_2^* \Pi_1^* c' \cdot g_{2*}a_*s(\mathcal{D})\}_{2x} \qquad \text{by (FP) and } (\mathcal{R}_3)$$
$$= \{i_2^* \Pi_1^* c' \cdot i_2^* s(F)\}^{z-x} \quad (\text{ in } \Pi_2^{-1}(X), \text{ dim} = 2x \text{ is equivalent to codim} = z - x)$$
$$= i_2^*\{\Pi_1^* c' \cdot s(F)\}^{z-x} \quad .$$

A fortiori in $CH_{2x}(\widehat{H^2}(Z)) = CH^{2z-2x}(\widehat{H^2}(Z))$, one gets :

$$i_{2*}g_{2*}\{cN \cdot a_*s(\mathcal{D})\}_{2x} = i_{2*}i_2^*\{\Pi_1^* c' \cdot s(F)\}^{z-x}$$
$$= \{\Pi_1^* c' \cdot s(F)\}^{z-x} \cdot [\Pi_2^{-1}(X)] \quad , \text{ by (FP)}.$$

Finally the formula in $CH^{2z-2x}(\widehat{H^2}(Z))$

$$[\widehat{H^2}(X)] = [\Pi_1^{-1}(X)] \cdot [\Pi_2^{-1}(X)] - \{\Pi_1^* c' \cdot s(F)\}^{z-x} \cdot [\Pi_2^{-1}(X)]$$

follows.

But for all cycles α with support in F, one has $\alpha \cdot [\Pi_1^{-1}(X)] = \alpha \cdot [\Pi_2^{-1}(X)]$ in $CH^{\bullet}(\widehat{H^2}(Z))$; one applies this to $\alpha = \{\Pi_1^* c' \cdot s(F)\}^{z-x}$. Moreover, equation (2.1) yields $[\Pi_i^{-1}(X)] = \Pi_i^*[X]$.

   Therefore we have shown the following theorem (Ran [Ra], p. 90, with $k = 2$) :

Theorem 1 (Ran) *Let $j : X \hookrightarrow Z$ be the canonical imbedding of a smooth subvariety of dimension x in a smooth variety of dimension z. Assume that the total Chern class $cv(X, Z)$ of the normal bundle can be written as $j^* c'$, where $c' \in CH^{\bullet}(Z)$.*

*   If Π_1, Π_2 : $\widehat{H^2}(Z) \to Z$ are the two natural morphisms, then one has in $CH^{\bullet}(\widehat{H^2}(Z))$ the equality :*

$$[\widehat{H^2}(X)] = \Pi_1^*[X] \cdot (\Pi_2^*[X] - \{s(F) \cdot \Pi_1^* c'\}^{z-x})$$

where $s(F)$ denotes the Segre class $s(F, \widehat{H^2}(Z))$ of the exceptional divisor.

2.2   Definition of the double class

From now on, let $f{:}V \longrightarrow W$ be a morphism of fixed proper smooth varieties, of respective dimensions n and m. Let $k = m - n = \text{cod}(f)$.

Hypothesis 2 :

(i) Assume $m > n$ (i.e. $k > 0$).

(ii) Assume $2n - m \geq 0$ (i.e. $k \leq n$) , so that the classes defined below are meaningful.

Let Γ be the graph of f and $j : \Gamma \hookrightarrow V \times W$ be the canonical imbedding. As in § 2.1, $\widehat{H^2(\Gamma)}$ can be identified with a subvariety of codimension $2m$ of $H^2(\widehat{V \times W})$.

Moreover, one has an imbedding

$$\alpha : \widehat{H^2(V)} \times W \hookrightarrow H^2(\widehat{V \times W})$$

defined canonically. Thus, if $d \subset V$ is a doublet and w a point of W, then $\alpha(d, w)$ denotes the doublet image of d by the imbedding

$$\begin{array}{ccc} V & \hookrightarrow & V \times W \\ v & \mapsto & (v, w) \end{array} \quad .$$

These doublets might be called "horizontal".

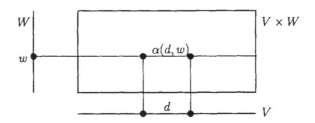

Let us denote by $P'_1 : \widehat{H^2(V)} \times W \to V$ the morphism $((v_1, d), w) \mapsto v_1$. The diagram

$$\widehat{H^2(\Gamma)} \hookrightarrow H^2(\widehat{V \times W}) \xleftarrow{\alpha} \widehat{H^2(V)} \times W \xrightarrow{P'_1} V$$

follows.

Definition 5 : Using the preceding notation, let :

(a) $M_2 = \alpha^*[\widehat{H^2(\Gamma)}] \quad \in \quad CH_{2n-m}(\widehat{H^2(V)} \times W)$

(b) $m_2 = P'_{1*}M_2 \quad \in \quad CH_{2n-m}(V) = CH^k(V)$.

One says that m_2 is the "double class".

Remark 1 : The cycle M_2 corresponds to the idea of points v_1, v_2 of V having the same image by f ; while the cycle m_2 corresponds to the idea of points $v_1 \in V$ such that there exists $v_2 \in V$ having the same image as v_1.

The question is to evaluate m_2 in $\mathbf{CH}^k(V)$. To do so, one introduces some new notation :

Notation 4 :

(i) If $g : X \longrightarrow Y$ is a morphism, one denotes by $\tilde{g}$ and g' the two following morphisms :
$$\begin{aligned} \tilde{g} &= g \times id_W : X \times W \longrightarrow Y \times W \\ g' &= g \times id_V : X \times V \longrightarrow Y \times V \end{aligned}$$

(ii) One denotes by pr_1 and pr_2 the projections from $V \times W$ onto V and W, respectively.

(iii) One denotes by p_1 and p_2 the two projections from $V \times V$ onto V. Note that
$$\bar{p}_1 = id_V \times pr_2 : V \times V \times W \longrightarrow V \times W \quad .$$

(iv) Let us denote by $\pi_i : \widehat{H^2(V)} \longrightarrow V$ the morphism which sends $((v_1, v_2), d)$ onto v_i, for $i = 1, 2$. Then one denotes by $\pi : \widehat{H^2(V)} \longrightarrow V \times V$ the morphism (π_1, π_2). Note that π can be identified with the blowing-up of the diagonal of $V \times V$.

(v) One denotes by $p_W : \widehat{H^2(V)} \times W \longrightarrow W$ the natural projection.

(vi) Last, as said above, if Γ is the graph of f, one denotes by $j : \Gamma \hookrightarrow V \times W$ the canonical imbedding. One has inverse isomorphisms :
$$\gamma : V \xrightarrow{\sim} \Gamma \quad \text{and} \quad \sigma : \Gamma \xrightarrow{\sim} V \quad .$$

With this notation, one has a commutative diagram :

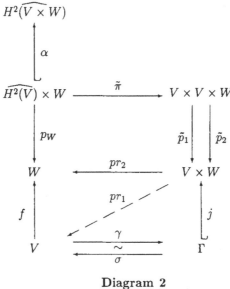

Diagram 2

Warning : We have represented pr_1 with a dotted line, since pr_1 <u>does not make the diagram commute</u>! In fact, $f \circ pr_1 \neq pr_2$; however $pr_1 \circ j = \sigma$.

Remark 2 : With this notation, one has :

$$m_2 = pr_{1*}\tilde{p}_{1*}\tilde{\pi}_* M_2 \text{ in } CH^k(V) \quad .$$

2.3 Computation of the double class

2.3.1 Computation of M_2

Let us apply theorem 1 with $X = \Gamma$, $Z = V \times W$ and $c' = pr_2^* cW$, where cW denotes the total Chern class of the tangent bundle TW to W. Indeed, the normal bundle to the graph is identified with $j^* pr_2^* TW$. For $x = n$, $z = n+m$, the formula of theorem 1 gives in $CH^\bullet(H^2(\widehat{V \times W}))$:

$$[\widehat{H^2(\Gamma)}] = \Pi_1^*[\Gamma] \cdot (\Pi_2^*[\Gamma] - \{s(\mathbb{F}) \cdot \Pi_1^* c'\}^m)$$

where $\mathbb{F} \subset H^2(\widehat{V \times W})$ denotes the exceptional divisor.

But one has the commutative diagram :

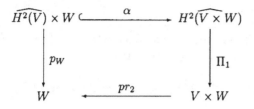

It follows that :

$$\alpha^* \Pi_1^* c' = \alpha^* \Pi_1^* pr_2^* cW = p_W^* cW = [\widehat{H^2(V)}] \times cW \quad .$$

Moreover, one has the commutative diagram :

$$
\begin{array}{ccc}
\widehat{H^2(V)} \times W & \xrightarrow{\ \ \alpha\ \ } & H^2(\widehat{V \times W}) \\[1ex]
\Big\downarrow{\tilde{\pi}} & & \Big\downarrow{\Pi_i} \qquad i = 1, 2 \\[1ex]
V \times V \times W & \xrightarrow{\ \tilde{p}_i\ } & V \times W
\end{array}
$$

hence $\alpha^* \Pi_i^* = \tilde{\pi}^* \tilde{p}_i^*$. Also,

$$\alpha^*[\mathbb{F}] = [F] \times [W]$$

where F denotes in this case the exceptional divisor of $\widehat{H^2(V)}$. (This comes from the fact that $\mathbb{F}$ and $\widehat{H^2(V)} \times W$ intersect transversally in $H^2(\widehat{V \times W})$, as shown by a computation on coordinates (4.32) performed in § 4.10).

It finally leads to the equality in $\mathrm{CH}^\bullet(H^2(\widehat{V \times W}))$:

$$M_2 = \alpha^*[\widehat{H^2(\Gamma)}] = \tilde{\pi}^* \tilde{p}_1^*[\Gamma] \cdot (\tilde{\pi}^* \tilde{p}_2^*[\Gamma] - \{s(F) \times cW\}^m) \qquad (2.2)$$

Note that $\alpha^* s(\mathbb{F}) = s(F)$ since the Segre classes are the inverse of the total Chern classes of the normal bundles.

2.3.2

Let us apply the projection formula to the morphism $\tilde{\pi}$:

$$\tilde{\pi}_* M_2 = \tilde{p}_1^*[\Gamma] \cdot \tilde{\pi}_*(\tilde{\pi}^* \tilde{p}_2^*[\Gamma] - \{s(F) \times cW\}^m) .$$

Since $\tilde{\pi}$ is birational, $\tilde{\pi}_* 1 = 1$. Hence, since $\tilde{\pi} = \pi \times id_W$ and from (FP) :

$$\tilde{\pi}_* M_2 = \tilde{p}_1^*[\Gamma] \cdot (\tilde{p}_2^*[\Gamma] - \{\pi_* s(F) \times cW\}^m) .$$

(we have used $(\mathcal{R}_2)$ with $\mathrm{cod}(\tilde{\pi}) = 0$).

But $\pi^{-1}(\Delta_V) = F$ where $\Delta_V \subset V \times V$ denotes the diagonal. From [FU2], proposition 4.2.b, one has $\pi_* s(F) = s(\Delta_V)$. Hence the equality in $\mathrm{CH}^\bullet(V \times V \times W)$

$$\tilde{\pi}_* M_2 = \tilde{p}_1^*[\Gamma] \cdot (\tilde{p}_2^*[\Gamma] - \{s(\Delta_V) \times cW\}^m)$$

holds.

This time let us apply the projection formula to the morphism $\tilde{p}_1$. Since $\text{cod}(\tilde{p}_1) = -n$, we get by $(\mathcal{R}_2)$ the equality in $\text{CH}^\bullet(V \times W)$

$$\tilde{p}_{1*}\tilde{\pi}_* M_2 = [\Gamma] \cdot \left(\tilde{p}_{1*}\tilde{p}_2^*[\Gamma] - \{p_{1*}s(\Delta_V) \times cW\}^{m-n}\right).$$

But the commutative diagram

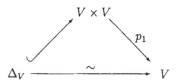

yields $p_{1*}s(\Delta_V) = c(V)^{-1}$, the inverse total Chern class of V, since the normal bundle to Δ_V can be identified with the tangent bundle to V.

Notation 5 : One writes $\bar{c} = c(V)^{-1} \times cW \in \text{CH}^\bullet(V \times W)$.

Notation 6 : Let $\mu = \mu_1 - \mu_2$ in $\text{CH}^k(V \times W)$,
where $\mu_1 = \tilde{p}_{1*}\tilde{p}_2^*[\Gamma]$ and $\mu_2 = \{\bar{c}\}^k$.

Using the previous notation and the fact that $m - n = k$, one gets :

$$\tilde{p}_{1*}\tilde{\pi}_* M_2 = [\Gamma] \cdot \left(\tilde{p}_{1*}\tilde{p}_2^*[\Gamma] - \{\bar{c}\}^k\right) = [\Gamma] \cdot \mu \quad . \tag{2.3}$$

From remark 2, one has in $\text{CH}^k(V)$: $m_2 = pr_{1*}\tilde{p}_{1*}\tilde{\pi}_* M_2 = pr_{1*}([\Gamma] \cdot \mu)$. This is also (from (FP) applied to j) : $m_2 = pr_{1*}(j_*j^*\mu) = \gamma^*j^*\mu$ car $pr_{1*}j_* = \sigma_* = \gamma^*$. Hence :

$$m_2 = \gamma^*j^*\mu \quad \text{in} \quad \text{CH}^k(V) \quad . \tag{2.4}$$

We will need the following lemma :

Lemma 2 *Let $\alpha \in CH^\bullet(V \times W)$. Then one has the equality $\tilde{p}_{1*}\tilde{p}_2^*\alpha = pr_2^* pr_{2*}\alpha$.*

Proof :

By the construction of $\tilde{p}_2$, one has $\tilde{p}_2^*\alpha = [V] \times \alpha \in \text{CH}^\bullet(V \times V \times W)$. Also, $\tilde{p}_1 = id_V \times pr_2$. Hence $\tilde{p}_{1*}\tilde{p}_2^*\alpha = [V] \times pr_{2*}\alpha = pr_2^* pr_{2*}\alpha$. $\qquad\square$

2.3.3

We saw (equation (2.4)) that $m_2 = \gamma^*j^*\mu = \gamma^*j^*\mu_1 - \gamma^*j^*\mu_2$ (using notation 6).
Let us first compute $\gamma^*j^*\mu_1$, i.e. $\boldsymbol{\gamma^*j^*\tilde{p}_{1*}\tilde{p}_2^*[\Gamma]}$.

From the previous lemma, $\gamma^*j^*\mu_1 = \gamma^*j^*pr_2^*pr_{2*}[\Gamma] = f^*pr_{2*}[\Gamma]$, since $f = pr_2 \circ j \circ \gamma$. Moreover, since $f_* = pr_{2*}j_*\gamma_*$, one has :

$$pr_{2*}[\Gamma] = f_*[V] . \tag{2.5}$$

Therefore, we have shown that :

$$\gamma^*j^*\mu_1 = f^*f_*[V] \quad \text{in} \quad CH^k(V) . \tag{2.6}$$

Let us now compute $\gamma^*j^*\mu_2$, i.e. $\gamma^*j^*\{\bar{c}\}^k$.
We have let $\bar{c} = c(V)^{-1} \times cW$ in $CH^\bullet(V \times W)$. But it can also be written as $\bar{c} = pr_1^*c(V)^{-1} \cdot pr_2^*cW$. Hence $j^*\bar{c} = j^*pr_1^*c(V)^{-1} \cdot j^*pr_2^*cW$.

But $pr_1 \circ j = \sigma$ and $pr_2 \circ j = f \circ \sigma$. Hence $j^*\bar{c} = \sigma^*(c(V)^{-1} \cdot f^*cW)$, and consequently :

$$j^*\{\bar{c}\}^k = \sigma^*\{c(V)^{-1} \cdot f^*cW\}^k$$

Notation 7 : In the Grothendieck's group $K(V)$, let us denote by

$$\nu(f) = f^*TW - TV ,$$

the <u>virtual normal bundle</u> to f. One denotes by c_i the Chern classes of $\nu(f)$.

Since $c\nu(f) = f^*cW \cdot c(V)^{-1}$, one gets :

$$j^*\{\bar{c}\}^k = \sigma^*c_k . \tag{2.7}$$

Finally, $\gamma^*\sigma^* = identity$ leads to

$$\gamma^*j^*\mu_2 = \gamma^*j^*\{\bar{c}\}^k = \gamma^*\sigma^*c_k = c_k . \tag{2.8}$$

It follows that :

$$m_2 = \gamma^*j^*(\mu_1 - \mu_2) = f^*f_*[V] - c_k .$$

We have just proved again the following theorem :

Theorem 2 (Laksov) Let $f : V \longrightarrow W$ be a morphism of proper and smooth varieties, with $\dim W = k + \dim V$ ($k > 0$). Let $m_2 \in CH^k(V)$ be the "double class", direct image of

$$M_2 = [\widehat{H^2(\Gamma)}] \cdot [\widehat{H^2(V)} \times W] \quad \in \quad CH^\bullet(H^2\widehat{(V \times W)})$$

where Γ is the graph of f. Then one has the <u>double formula</u> :

$$m_2 = f^*f_*[V] - c_k$$

where c_k is the kth Chern class of the virtual normal bundle $f^*TW - TV$.

Chapter 3

Triple formula

3.1 The class of $\widehat{H^3(X)}$ in $\widehat{H^3(Z)}$

Let Z be a smooth variety of dimension z. One introduces in [LB1] the smooth variety $\widehat{H^3(Z)}$, of dimension $3z$, of "complete triples". Below we recall briefly its construction.

Notation 8 : Let us denote by $H^i(Z)$ the Hilbert scheme $Hilb^i(Z)$ (cf. [G], [I1]). A complete triple of Z consists of the data $\hat{t} = (p_1, p_2, p_3, d_{12}, d_{23}, d_{31}, t) \in Z^3 \times (H^2(Z))^3 \times H^3(Z)$ where

$$\begin{cases} p_i \subset d_{ij} \subset t & (\text{ scheme-theoretic inclusions }) \\ p_j = Res(p_i, d_{ij}) \\ p_k = Res(d_{ij}, t) & \text{with } \{i, j, k\} = \{1, 2, 3\} \end{cases}$$

Let us denote by $\widehat{H^3(Z)}$ the set of complete triples of Z. We show that it is a nonsingular variety, birational to $Z \times Z \times Z$. In the case where $Z = \mathbb{P}^2$, the variety $\widehat{H^3(\mathbb{P}^2)}$ is canonically isomorphic to the Semple's complete triangles variety [S], since studied by Roberts-Speiser [RS1].

One has in $\widehat{H^3(Z)}$ for Cartier's divisors :

$$\begin{cases} E_{12} = \{\, \hat{t} \mid d_{23} = d_{31} \} \\ E_{23} = \{\, \hat{t} \mid d_{12} = d_{31} \} \qquad \text{and} \quad E^\bullet = \{\, \hat{t} \mid t \text{ amorphous} \} \\ E_{31} = \{\, \hat{t} \mid d_{23} = d_{12} \} \end{cases}$$

Recall that (cf. definition 2) an <u>amorphous</u> triplet t is such that $supp(t)$ is reduced to only one point p and t is defined by the square of the ideal of p in a germ of a smooth surface containing p.

Then let $j : X \hookrightarrow Z$ be the canonical imbedding of a smooth subvariety of dimension x. The variety $\widehat{H^3(X)}$ can clearly be identified with a subvariety of $\widehat{H^3(Z)}$, as one sees by coordinate computations (see (4.30) in § 4.9).

The question is to evaluate the fundamental class of $[\widehat{H^3(X)}] \in CH^\bullet(\widehat{H^3(Z)})$, with the same hypothesis as in § 2.1, i.e. $cv(X, Z) = j^*c'$ where $c' \in CH^\bullet(Z)$.

Let

$$P_i : \widehat{H^3(Z)} \to Z$$
$$\hat{t} \mapsto p_i$$

be the three natural morphisms. A computation (see (4.25) in § 4.7) shows that the P_i are flat. Similarly, one has the morphism

$$P_{12} : \widehat{H^3(Z)} \to \widehat{H^2(Z)}$$
$$\hat{t} \mapsto (p_1, d_{12})$$

and a computation (see (4.23)) shows that P_{12} is flat.

Then the intersection $I = P_{12}^{-1}(\widehat{H^2(X)}) \cap P_3^{-1}(X)$ in $\widehat{H^3(Z)}$ possesses three excess components, respectively

$$\mathcal{E}_{23} = E_{23} \cap P_{12}^{-1}(\widehat{H^2(X)}), \quad \mathcal{E}_{31} = E_{31} \cap P_{12}^{-1}(\widehat{H^2(X)}), \quad \mathcal{E}^\bullet = E^\bullet \cap P_{12}^{-1}(\widehat{H^2(X)}).$$

In the following drawing, a doublet of support one point (therefore isomorphic to $Spec(\mathbb{C}[T]/(T^2))$) is represented by :

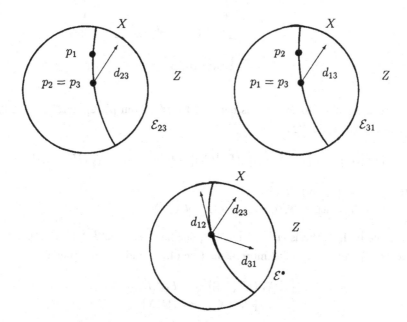

Note that in the last drawing, d_{12} is drawn "tangent" to X, because we have the scheme-theoretic inclusion $d_{12} \subset X$.

Notation 9 : The sum of the three divisors $E_{23} + E_{31} + E^{\bullet}$ of $\widehat{H^3(Z)}$ is denoted by $\overline{E}$ and $\mathcal{E}_{23} + \mathcal{E}_{31} + \mathcal{E}^{\bullet} \subset I$ by $\overline{\mathcal{E}}$.

Thus one has the commutative diagram where the parentheses indicate the dimensions and where the imbeddings are the canonical imbeddings :

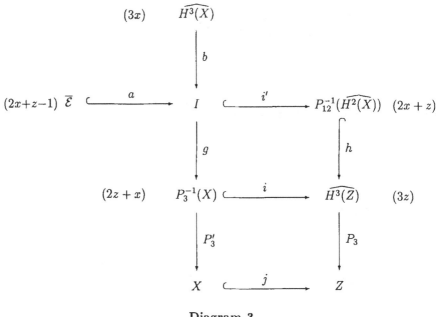

Diagram 3

The restriction of P_3 to $P_3^{-1}(X)$ is denoted by P_3'. From [FU2], corollary 9.2.2, one has the equality in $CH^{\bullet}(I)$:

$$[\widehat{H^3(X)}] = [P_{12}^{-1}(\widehat{H^2(X)})] \cdot [P_3^{-1}(X)] - \{cN \cdot a_* s(\overline{\mathcal{E}}, P_{12}^{-1}(\widehat{H^2(X)}))\}_{3x}$$

where $N = g^* \nu(P_3^{-1}(X), \widehat{H^3(Z)})$.

But P_3 is flat (see (4.25)), so $N = g^* P_3'^* \nu(X, Z)$.

Then, since by hypothesis $c\nu(X, Z) = j^* c'$, one has : $cN = g^* P_3'^* j^* c' = i'^* h^* P_3^* c'$.
Moreover, P_{12} and P_3 are flat morphisms (see (4.23) and (4.25)), therefore

$$\begin{cases} [P_{12}^{-1}(\widehat{H^2(X)})] & = & P_{12}^*[\widehat{H^2(X)}] \\ [P_3^{-1}(X)] & = & P_3^*[X] \end{cases} \tag{3.1}$$

Moreover, one has the following lemma

Lemma 3 *The intersection* $\overline{\mathcal{E}} = \overline{E} \cap P_{12}^{-1}(\widehat{H^2(X)})$ *is transverse in* $\widehat{H^3(Z)}$.

Proof :

See (4.31) in § 4.9. □

This implies the equality of Segre classes (inverse of Chern classes) :

$$i'_* a_* s(\overline{\mathcal{E}}, P_{12}^{-1}(\widehat{H^2(X)})) = h^* s(\overline{E}, \widehat{H^3(Z)}) .$$

Notation 10 : One writes $s(\overline{E})$ for $s(\overline{E}, \widehat{H^3(Z)})$ and also $s(\overline{\mathcal{E}})$ for $s(\overline{\mathcal{E}}, P_{12}^{-1}(\widehat{H^2(X)}))$.

Using what is above, one writes in $\mathrm{CH}_{3x}(P_{12}^{-1}(\widehat{H^2(X)}))$:

$$
\begin{aligned}
i'_* \{cN \cdot a_* s(\overline{\mathcal{E}})\}_{3x} &= i'_* \{i'^* h^* P_3^* c' \cdot a_* s(\overline{\mathcal{E}})\}_{3x} \\
&= \{h^* P_3^* c' \cdot i'_* a_* s(\overline{\mathcal{E}})\}_{3x} \qquad \text{from} \quad (\mathrm{FP}) \quad \text{and} \quad (\mathcal{R}_3) \\
&= \{h^* P_3^* c' \cdot h^* s(\overline{E})\}^{z-x} (\text{in } P_{12}^{-1}(\widehat{H^2(X)}), \text{ the dimension } 3x \\
&\quad \text{ is equivalent to the codimension } z - x.) \\
&= h^* \{P_3^* c' \cdot s(\overline{E})\}^{z-x} .
\end{aligned}
$$

A fortiori, we get in $\mathrm{CH}_{3x}(\widehat{H^3(Z)}) = \mathrm{CH}^{3z-3x}(\widehat{H^3(Z)})$:

$$
\begin{aligned}
h_* i'_* \{cN \cdot a_* s(\overline{E})\}_{3x} &= h_* h^* \{P_3^* c' \cdot s(\overline{E})\}^{z-x} \\
&= \{P_3^* c' \cdot s(\overline{E})\}^{z-x} \cdot [P_{12}^{-1}(\widehat{H^2(X)})] \qquad \text{from} \quad (\mathrm{FP}).
\end{aligned}
$$

Finally the formula in $\mathrm{CH}^{3z-3x}(\widehat{H^3(Z)})$

$$[\widehat{H^3(X)}] = [P_{12}^{-1}(\widehat{H^2(X)})] \cdot ([P_3^{-1}(X)] - \{P_3^* c' \cdot s(\overline{E})\}^{z-x})$$

follows.

But we noticed that P_{12} and P_3 are flat ; with the help of (3.1), we have therefore proved the following theorem (which generalizes Ran [Ra], p. 90 for $k = 3$) :

Theorem 3 *Let $j : X \hookrightarrow Z$ be the canonical imbedding of a smooth subvariety of dimension x in a smooth variety of dimension z. Suppose that the total Chern class $c\nu(X, Z)$ of the normal bundle can be written as $j^* c'$ where $c' \in CH^\bullet(Z)$. Let P_{12} and P_3 be the morphisms defined in the following way : if $\hat{t} = (p_1, p_2, p_3, d_{12}, d_{23}, d_{31}, t)$ is a complete triplet of Z, let*

$$
\begin{array}{ccccc}
P_{12} : & \widehat{H^3(Z)} & \to & \widehat{H^2(Z)} & \text{and} \qquad P_3 : & \widehat{H^3(Z)} & \to & Z \\
& \hat{t} & \mapsto & (p_1, d_{12}) & & \hat{t} & \mapsto & p_3 .
\end{array}
$$

Then the equality in $CH^\bullet(\widehat{H^3(Z)})$ follows :

$$[\widehat{H^3(X)}] = P_{12}^*[\widehat{H^2(X)}] \cdot (P_3^*[X] - \{s(\overline{E}) \cdot P_3^* c'\}^{z-x}) \qquad (3.2)$$

where $s(\overline{E})$ denotes the Segre class $s(\overline{E}, \widehat{H^3(Z)})$ of the divisor $\overline{E} = E_{23} + E_{31} + E^\bullet$ of $\widehat{H^3(Z)}$.

3.2   The triple formula

3.2.1   Some notation

The notation of § 2.2 for a morphism $f : V \longrightarrow W$ of smooth varieties is used again :
$n = \dim V$, $m = \dim W$ and $m = k + n$, $k > 0$.

Hypothesis 3 : Suppose that $3n - 2m \geq 0$, i.e. $k \leq n/2$, so that the classes defined
below are meaningful.

One has a natural imbedding

$$\beta : \widehat{H^3(V)} \times W \hookrightarrow H^3(\widehat{V \times W})$$

similar to the imbedding $\alpha : \widehat{H^2(V)} \times W \hookrightarrow H^2(\widehat{V \times W})$ seen in § 2.2. Its image
consists of the "horizontal" triplets of $V \times W$.

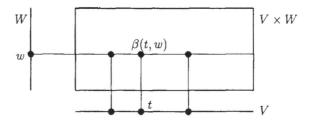

Let us denote by $P_1'' : \widehat{H^3(V)} \times W \longrightarrow V$ the morphism which takes $(\hat{t}, w)$ to v_1,
where $\hat{t} = (t, d_{12}, d_{23}, d_{31}, v_1, v_2, v_3)$ is in $\widehat{H^3(V)}$.

If Γ is the graph of f, the diagram

$$\widehat{H^3(\Gamma)} \hookrightarrow H^3(\widehat{V \times W}) \overset{\beta}{\leftarrow} \widehat{H^3(V)} \times W \overset{P_1''}{\rightarrow} V$$

follows.

Definition 6 : With the above notation, let

(a) $M_3 = \beta^*[\widehat{H^3(\Gamma)}] \quad \in \quad \mathrm{CH}_{3n-2m}(\widehat{H^3(V)} \times W)$

(b) $\widehat{m_3} = P_1''{}_* M_3 \quad \in \quad \mathrm{CH}_{3n-2m}(V) = \mathrm{CH}^{2k}(V)$.

$\widehat{m_3}$ is said to be the "triple class".

Remark 3 : As in remark 2, the cycle M_3 corresponds intuitively to points v_1, v_2,
v_3 of V having the same image by f; while the cycle $\widehat{m_3}$ corresponds intuitively to
points $v_1 \in V$ such that there exists v_2, v_3 in V having the same image as v_1.

The question is to evaluate $\widetilde{m_3}$ in $\mathbf{CH}^{2k}(V)$. To do so, we introduce some additional notation besides the morphisms considered in § 2.2 :

Notation 11 :

(i) Let $q : \widehat{H^2(V)} \times V \longrightarrow \widehat{H^2(V)}$ be the natural projection ;

(ii) Let
$$
\begin{array}{cccc}
\omega_i : & \widehat{H^2(V)} \times V & \to & V \qquad i = 1, 2, 3 \\
& (\hat{d}, v_3) & \mapsto & v_i
\end{array}
$$

where $\hat{d} = (v_1, v_2, d) \in \widehat{H^2(V)}$. Note the asymmetric role of 3 in comparison with 1 and 2 in this notation.

(iii) Let us denote by $\bar{p}_W : \widehat{H^3(V)} \times W \longrightarrow W$ the natural projection.

(iv) Let ϕ be the birational morphism
$$
\begin{array}{cccc}
\phi : & \widehat{H^3(V)} & \to & \widehat{H^2(V)} \times V \\
& \hat{t} & \mapsto & ((v_1, v_2, d_{12}), v_3)
\end{array}
$$

if $\hat{t} = (t, d_{12}, d_{23}, d_{31}, v_1, v_2, v_3)$.

(v) The imbedding $\beta : \widehat{H^3(V)} \times W \hookrightarrow H^3(\widehat{V \times W})$ has been defined above.

(vi) Recall (see notation 4.(i)) that $\bar{g}$ denotes $g \times id_W$, where g is a morphism.

With this notation (in addition to the notation of § 2.2), one has a commutative diagram :

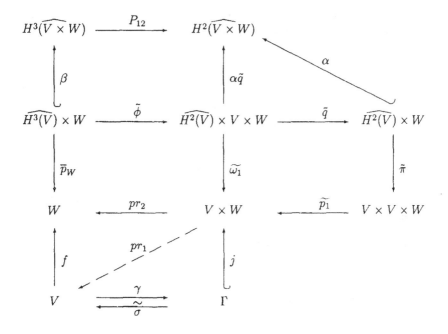

Diagram 4

Warning: pr_1 is represented by a dotted line, since it does not make the diagram commute ; however, $pr_1 \circ j = \sigma$.

Remark 4 : With this notation, one has $\widehat{m_3} = pr_{1*}\widetilde{\omega}_{1*}\widetilde{\phi}_*M_3 \in \mathrm{CH}^{2k}(V)$.

3.2.2 Computation of M_3 and $\widetilde{\phi}_*M_3$

In the same way as in § 2.3 with theorem 1, one applies theorem 3 with $X = \Gamma$, $Z = V \times W$ and $c' = pr_2^*cW$. Theorem 3 yields the equality in $\mathrm{CH}^{\bullet}(H^3\widehat{(V \times W)})$:

$$[\widehat{H^3(\Gamma)}] = P_{12}^*[\widehat{H^2(\Gamma)}] \cdot (P_3^*[\Gamma] - \{s(\overline{\overline{\mathbb{E}}}) \cdot P_3^*c'\}^m) \ ,$$

where $\overline{\overline{\mathbb{E}}}$ is $\mathbb{E}_{23} + \mathbb{E}_{31} + \mathbb{E}^{\bullet}$. The double bar denotes the divisors related to $H^3\widehat{(V \times W)}$. The divisors related to $\widehat{H^3(V)}$ are denoted by $\overline{E} = E_{23} + E_{31} + E^{\bullet}$, so that $\beta^*(\overline{\mathbb{E}}) = \overline{E} \times [W]$ by transversality (cf. (4.33) in § 4.10). The definition of M_3 (see definition 6.(a)) yields :

$$M_3 = \beta^*P_{12}^*[\widehat{H^2(\Gamma)}] \cdot (\beta^*P_3^*[\Gamma] - \{(s(\overline{E}) \times [W]) \cdot \beta^*P_3^*c'\}^m) \quad .$$

On the other hand, as it can be checked easily, $pr_2 \circ P_3 \circ \beta = \overline{p}_W$. Therefore

$$\beta^*P_3^*c' = \beta^*P_3^*pr_2^*cW = \overline{p}_W^*cW = [\widehat{H^3(V)}] \times cW \ .$$

Furthermore, as it can be seen on the previous diagram 4, $\beta^* P_{12}^* = \tilde{\phi}^* \tilde{q}^* \alpha^*$. Moreover, one verifies easily that $P_3 \circ \beta = \widetilde{\omega_3} \circ \tilde{\phi}$. Therefore $\beta^* P_3^* = \tilde{\phi}^* \widetilde{\omega_3}^*$.

Finally, one has in $\mathrm{CH}^\bullet(\widehat{H^3(V)} \times W)$:

$$M_3 = \tilde{\phi}^* \tilde{q}^* \alpha^* [\widehat{H^2(\Gamma)}] \cdot (\tilde{\phi}^* \widetilde{\omega_3}^*[\Gamma] - \{s(\overline{E}) \times cW\}^m) \quad .$$

Let us then apply the projection formula to the birational morphism $\tilde{\phi} = \phi \times id_W$. Since $\mathrm{cod}(\tilde{\phi}) = 0$ and $\tilde{\phi}_* \tilde{\phi}^* = id$, one has from $(\mathcal{R}_2)$:

$$\tilde{\phi}_* M_3 = \tilde{q}^* \alpha^* [\widehat{H^2(\Gamma)}] \cdot (\widetilde{\omega_3}^*[\Gamma] - \{\phi_* s(\overline{E}) \times cW\}^m) \quad . \tag{3.3}$$

Let us then introduce some additional notation.

Notation 12 :

(a) Let $\Theta : U \longrightarrow H^2(V)$ be the universal two-sheeted covering and $R \subset U$ its ramification locus. One has $U \subset H^2(V) \times V$. Of course, it is the same as $\theta : \widehat{H^2(V)} \longrightarrow H^2(V)$, but we denote it differently in order to avoid any confusion.
Similarly, $R \subset U$ corresponds to $F \subset \widehat{H^2(V)}$ and it is the ramification locus of Θ (see § 1.3).

(b) Let $\theta' = \theta \times id_V$. One constructs the cartesian diagram (where θ'' is the restriction of θ' and $u, \hat{u}$ are the canonical imbeddings) :

$$
\begin{array}{ccc}
\hat{U} \,\hookrightarrow & \xrightarrow{\;\hat{u}\;} & \widehat{H^2(V)} \times V \\
\downarrow{\scriptstyle \theta''} & & \downarrow{\scriptstyle \theta'} \\
U \,\hookrightarrow & \xrightarrow{\;u\;} & H^2(V) \times V
\end{array}
$$

(c) Let G_{13} and G_{23} be the graphs in $\widehat{H^2(V)} \times V$ of the natural morphisms :

$$
\begin{array}{rccc}
P_i : & \widehat{H^2(V)} & \to & V \\
& ((v_1, v_2), d) & \mapsto & v_i \, .
\end{array}
$$

(d) Let $B \subset F \times V$ be the graph of $P_{1|F} : F \longrightarrow V$.

(e) We denote by $D \subset H^2(V)$ the set consisting of the doublets of support one point.

(f) The Segre class $s(\widehat{U}, \widehat{H^2(V)} \times V)$ (resp. $s(U, H^2(V) \times V)$) is denoted by $s(\widehat{U})$ (resp. $s(U)$).

See drawing in § 3.2.4, page 43.

Lemma 4

(i) One has the inverse scheme theoretic image $\quad \phi^{-1}(\widehat{U}) = \overline{E}$ in $\widehat{H^3(V)}$.

(ii) One has the equality of schemes $\quad \widehat{U} = G_{13} \cup G_{23}$ in $\widehat{H^2(V)} \times V$.

(iii) One has the equality of schemes $\quad B = G_{13} \cap G_{23}$ in $\widehat{H^2(V)} \times V$.

(iv) One has the equality $\quad \theta^*[D] = 2[F]$ in $CH^1(\widehat{H^2(V)})$.

(v) One also has the equality $\quad \theta'^*[R] = 2[B]$ in $CH^\bullet(\widehat{H^2(V)} \times V)$.

Proof:

The proof consists in computing coordinates ; the results are shown in chapter 4 (see resp. (4.28), (4.15), (4.13), (4.9) and (4.16)).   □

Lemma 4.(i) in conjunction with prop. 4.2.a, p. 74 in [FU2] shows that the direct image by ϕ of the Segre class $s(\overline{E})$ is given by :

$$\phi_* s(\overline{E}) \;=\; s(\widehat{U}) \quad \text{in} \quad \text{CH}^\bullet(\widehat{H^2(V)} \times V). \tag{3.4}$$

Equation (3.3) implies the final result in $\text{CH}^\bullet(\widehat{H^2(V)} \times V \times W)$:

$$\tilde{\phi}_* M_3 \;=\; \tilde{q}^* \alpha^* [\widehat{H^2(\Gamma)}] \cdot (\widetilde{\omega_3}^*[\Gamma] - \{s(\widehat{U}) \times cW\}^m). \tag{3.5}$$

Let us introduce some more notation.

Notation 13 : Let $\quad \tilde{\phi}_* M_3 = \nu = \nu_1 - \nu_2 \quad \in \quad \text{CH}^\bullet(\widehat{H^2(V)} \times V \times W)$
where
$$\nu_1 = \tilde{q}^* \alpha^* [\widehat{H^2(\Gamma)}] \cdot \widetilde{\omega_3}^*[\Gamma]$$
$$\nu_2 = \tilde{q}^* \alpha^* [\widehat{H^2(\Gamma)}] \cdot \{s(\widehat{U}) \times cW\}^m \quad .$$

Remark 4 implies :

$$\widehat{m_3} = pr_{1*} \widetilde{\omega_1}_* \nu \quad . \tag{3.6}$$

3.2.3   Computation of $pr_{1*}\widetilde{\omega}_{1*}\nu_1$

As it can be seen on diagram 4, one has $\widetilde{\omega}_1 = \widetilde{p_1} \circ \tilde{\pi} \circ \tilde{q}$; one first calculates $\widetilde{\omega}_{1*}\nu_1 = \widetilde{p_{1*}}\tilde{\pi}_*\tilde{q}_*\nu_1$.

The definition of ν_1 (notation 13) and the application of (FP) to $\tilde{q}$ yield :

$$\tilde{q}_*\nu_1 = \alpha^*[\widehat{H^2(\Gamma)}] \cdot \tilde{q}_*\widetilde{\omega}_{3*}[\Gamma] \quad .$$

But $\widetilde{\omega}_3 : \widehat{H^2(V)} \times V \times W \longrightarrow V \times W$ simply is the natural projection. Therefore :

$$\widetilde{\omega_3}^*[\Gamma] = [\widehat{H^2(V)}] \times [\Gamma] \quad \in \quad CH^\bullet(\widehat{H^2(V)} \times V \times W) \quad ,$$

and since $\tilde{q}$ is the natural projection $\widehat{H^2(V)} \times V \times W \longrightarrow \widehat{H^2(V)} \times W$, one has :

$$\tilde{q}_*\widetilde{\omega_3}^*[\Gamma] = [\widehat{H^2(V)}] \times pr_{2*}[\Gamma] \ .$$

But we have seen that $pr_{2*}[\Gamma] = f_*[V]$ (see equality (2.5)). Therefore $\tilde{q}_*\widetilde{\omega_3}^*[\Gamma]$ can also be written as :

$$[\widehat{H^2(V)}] \times f_*[V] = \tilde{\pi}^*\widetilde{p_1}^*([V] \times f_*[V]) = \tilde{\pi}^*\widetilde{p_1}^*pr_2^*f_*[V] \quad .$$

We finally obtain

$$\tilde{q}_*\nu_1 = \alpha^*[\widehat{H^2(\Gamma)}] \cdot \tilde{\pi}^*\widetilde{p_1}^*pr_2^*f_*[V] \quad .$$

Applying (FP) to $\widetilde{p_1} \circ \tilde{\pi}$ yields in $CH^\bullet(V \times W)$:

$$\widetilde{\omega}_{1*}\nu_1 = \widetilde{p_{1*}}\tilde{\pi}_*\tilde{q}_*\nu_1 = \widetilde{p_{1*}}\tilde{\pi}_*\alpha^*[\widehat{H^2(\Gamma)}] \cdot pr_2^*f_*[V] \ .$$

But we let (definition 5): $M_2 = \alpha^*[\widehat{H^2(\Gamma)}]$. The equalities (2.3) yield :

$$\widetilde{p_{1*}}\tilde{\pi}_*[M_2] = [\Gamma] \cdot \mu \quad \in \quad CH^\bullet(V \times W) \quad .$$

Therefore $\widetilde{\omega}_{1*}\nu_1 = [\Gamma] \cdot \mu \cdot pr_2^*f_*[V].$

But, for all $a \in CH^\bullet(V \times W)$, (FP) gives :

$$j_*j^*a = a \cdot [\Gamma] \quad . \tag{3.7}$$

Therefore $\widetilde{\omega}_{1*}\nu_1 = j_*j^*(\mu \cdot pr_2^*f_*[V]).$ Since $\sigma = pr_1 \circ j$, one has :

$$pr_{1*}\widetilde{\omega}_{1*}\nu_1 = \sigma_*j^*(\mu \cdot pr_2^*f_*[V]) \quad .$$

But $\gamma^* = \sigma_*$ (since γ (resp. σ) is the inverse isomorphism of σ (resp. γ)). It follows that :

$$pr_{1*}\widetilde{\omega}_{1*}\nu_1 = \gamma^*j^*\mu \cdot \gamma^*j^*pr_2^*f_*[V] \quad .$$

On one hand, $\gamma^*j^*\mu = m_2$ (see (2.4)) ; on the other hand, $f = pr_2 \circ j \circ \gamma$. Thus we have obtained what we were looking for :

$$pr_{1*}\widetilde{\omega}_{1*}\nu_1 = m_2 \cdot f^*f_*[V] \quad . \tag{3.8}$$

3.2.4   Computation of $\{s(\widehat{U}) \times cW\}^m$

We already mentioned in the introduction that the notation is abused in the following way : if $Y \subset Y'$ is a subvariety, the class in $\mathrm{CH}^{\bullet}(Y')$ is denoted by Y and not by $[Y]$.

$a)$ In the following calculations, we need to know

$$s(\widehat{U}) \times cW \quad \in \quad \mathrm{CH}^{\bullet}(\widehat{H^2(V)} \times V \times W) \quad,$$

or more exactly (see § 1.2) : $\hat{u}_* s(\widehat{U}) \times cW$,
where $\hat{u} : s(\widehat{U}) \hookrightarrow \widehat{H^2(V)} \times V$ is the canonical imbedding. But one has (notation 12) a commutative diagram :

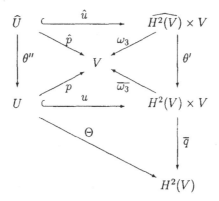

where w_3 and $\overline{w_3}$ are the natural projections on the second factor and $\hat{p}$, p are their restrictions ; $\overline{q}$ is the projection on the first factor. Θ is the two-sheeted universal covering. (We denote it by $\Theta : U \longrightarrow H^2(V)$ and not by $\theta : \widehat{H^2(V)} \longrightarrow H^2(V)$, since U is <u>another</u> copy of $\widehat{H^2(V)}$, in order to avoid any confusion. Same thing for $\overline{q}$ and q).

Notation 14 : In order to simplify the notation, one writes H for $H^2(V)$.

$b)$

Lemma 5 (Kleiman, Ronga) : Let ν be the normal bundle to U in $H \times V$. Then one has in $\mathrm{CH}^{\bullet}(U)$ the total Chern class :

$$c(\nu) = p^* c(TV) \cdot \frac{1 + 2R}{1 + R}$$

where TV is the tangent bundle to V and R is the ramification locus of U on H.

Proof :

From [H], II, prop. 8.12, one has the exact sequence of sheaves on U :

$$I/I^2 \longrightarrow \Omega^1_{U/H} \otimes \mathcal{O}_R \longrightarrow \Omega^1_{R/H} \longrightarrow 0$$

where $I = \mathcal{O}_U(-R)$ is the ideal of R. Then

$$I/I^2 = I \otimes \mathcal{O}_U/I \simeq \mathcal{O}_R(-R) .$$

Furthermore, (see (4.10) in § 4.3), the projection $R \longrightarrow H$ is not ramified, i.e. $\Omega^1_{R/H} = 0$. Thus one obtains the isomorphism $\Omega^1_{U/H} \simeq \mathcal{O}_R(-R)$. Let us apply again [H], II, prop. 8.12 but this time to the morphism $\bar{q}$. One obtains the exact sequence of sheaves on U :

$$0 \longrightarrow \nu^* \longrightarrow \bar{q}^*\Omega^1_V \mid U \longrightarrow \Omega^1_{U/H} \longrightarrow 0$$

ν^* denoting the conormal bundle. It follows that in the Grothendieck group $K(U)$ of coherent sheaves (or vector bundles), one has the equality :

$$\nu^* = p^*\Omega^1_V - \mathcal{O}_R(-R) .$$

From the exact sequence of sheaves on U :

$$0 \longrightarrow \mathcal{O}_U(-2R) \longrightarrow \mathcal{O}_U(-R) \longrightarrow \mathcal{O}_R(-R) \longrightarrow 0$$

one obtains in $K(U)$:

$$\nu^* = p^*\Omega^1_V - \mathcal{O}_U(-R) + \mathcal{O}_U(-2R) .$$

The equality of Chern polynomials follows :

$$C_t(\nu^*) = p^*C_t(\Omega^1_V) \cdot \frac{1 - 2tR}{1 - tR}$$

For $t = -1$, it yields the equality between total Chern classes of duals :

$$c(\nu) = p^*c(TV) \cdot \frac{1 + 2R}{1 + R} \quad \in \quad \mathrm{CH}^\bullet(U) .$$

Thus, lemma 5 has been shown.

c) From lemma 5, we get inverse Chern classes (by denoting $c(TV)$ by $c(V)$) :

$$c(\nu)^{-1} = p^*c(V)^{-1} \cdot \frac{1 + R}{1 + 2R} \quad \in \quad \mathrm{CH}^\bullet(U) .$$

But $\dfrac{1 + R}{1 + 2R} = 1 - R(1 + 2R)^{-1}$. Lemma 4.(iv) gives $\theta^*D = 2F$ in $\mathrm{CH}^1(\widehat{H^2(V)})$; which gives $\Theta^*D = 2R$ in $\mathrm{CH}^1(U)$, if rewritten in the other copy U of $\widehat{H^2(V)}$. Thus :

$$\frac{1 + R}{1 + 2R} = 1 - Ru^*\bar{q}^*(1 + D)^{-1} \quad \in \quad \mathrm{CH}^\bullet(U) .$$

One denotes (notation 12.(f)) by $s(U) \in \mathrm{CH}^\bullet(U)$ the Segre class of $U \subset H^2(V) \times V$. From ([FU2], chapter 4), one obtains :

$$s(U) = c(\nu)^{-1} = p^*c(V)^{-1} \cdot (1 - Ru^*\bar{q}^*(1 + D)^{-1}) \quad \in \quad \mathrm{CH}^\bullet(U) .$$

Let us apply (FP) to the morphism u; the equality in $\mathrm{CH}^\bullet(H^2(V) \times V)$

$$u_* s(U) = \overline{\omega_3}^* c(V)^{-1} \cdot (U - u_* R \cdot \overline{q}^* (1 + D)^{-1})$$

follows since $p^* = u^* \overline{\omega_3}^*$.

Let us lift by θ'. From ([FU2], prop. 4.2.b) the equality in $\mathrm{CH}^\bullet(\widehat{H^2(V)} \times V)$

$$\begin{aligned}
\hat{u}_* s(\widehat{U}) &= \hat{u}_* \theta''^* s(U) = \theta'^* u_* s(U) \\
&= \omega_3^* c(V)^{-1} \cdot (\widehat{U} - \theta'^* R \cdot \theta'^* \overline{q}^* (1 + D)^{-1}) \ .
\end{aligned}$$

follows since θ' is flat (θ is flat). But one has the commutative diagram (q and $\overline{q}$ are the two natural projections) :

$$
\begin{array}{ccc}
\widehat{H^2(V)} \times V & \xrightarrow{\ \ q\ \ } & \widehat{H^2(V)} \\
\downarrow{\theta'} & & \downarrow{\theta} \\
H^2(V) \times V & \xrightarrow{\ \ \overline{q}\ \ } & H^2(V)
\end{array}
$$

Thus, one has $\theta'^* \overline{q}^* D = q^* \theta^* D$; but $\theta^* D = 2F$ and $\theta''^* R = 2B$ (cf. lemma 4). Therefore, in $\mathrm{CH}^\bullet(\widehat{H^2(V)} \times V)$, one has :

$$\hat{u}_* s(\widehat{U}) = \omega_3^* c(V)^{-1} \cdot (\widehat{U} - 2B(1 + 2q^* F)^{-1}) \ . \tag{3.9}$$

d) Let us then look at $\{\hat{u}_* s(\widehat{U}) \times cW\}^m \in \mathrm{CH}^\bullet(\widehat{H^2(V)} \times V \times W)$. One can formally expand $\hat{u}_* s(\widehat{U})$ given by (3.9) to obtain :

$$\hat{u}_* s(\widehat{U}) = \omega_3^* c(V)^{-1} \cdot (\widehat{U} + B \sum_{h \geq 1} (-2)^h q^* F^{h-1}) \ .$$

Thus, the equality in $\mathrm{CH}^\bullet(\widehat{H^2(V)} \times V \times W)$

$$\{(\omega_3^* c(V)^{-1} \cdot \widehat{U}) \times cW\}^m + \sum_{h \geq 1} (-2)^h \{(\omega_3^* c(V)^{-1} \cdot Bq^* F^{h-1}) \times cW\}^m$$

follows.

For the first term in this sum, one has in $\mathrm{CH}^\bullet(\widehat{H^2(V)} \times V \times W)$:

$$(\omega_3^* c(V)^{-1} \cdot \widehat{U}) \times cW = (\omega_3^* c(V)^{-1} \times cW) \cdot (\widehat{U} \times W) \ .$$

But $\mathrm{codim}(\widehat{U} \times W) = n$. Applying $(\mathcal{R}_1)$ yields :

$$\begin{aligned}
\{(\omega_3^* c(V)^{-1} \cdot \widehat{U}) \times cW\}^m &= \{\omega_3^* c(V)^{-1} \times cW\}^{m-n} \cdot (\widehat{U} \times W) \\
&= \widetilde{\omega_3}^* \{\tilde{c}\}^k \cdot (\widehat{U} \times W)
\end{aligned}$$

(recall that $\bar{c} = c(V)^{-1} \times cW$ – see notation 5).

For the other terms of this sum, one has :

$$(\omega_3^* c(V)^{-1} \cdot Bq^* F^{h-1}) \times cW = (\omega_3^* c(V)^{-1} \times cW) \cdot ((Bq^* F^{h-1}) \times W) \ .$$

But $\text{codim}(Bq^* F^{h-1}) = (n+1) + (h-1) = n + h$. From $(\mathcal{R}_1)$ again, one has :

$$\begin{aligned}
\{(\omega_3^* c(V)^{-1} \cdot Bq^* F^{h-1}) \times cW\}^m &= \{\omega_3^* c(V)^{-1} \times cW\}^{m-n-h} \cdot ((Bq^* F^{h-1}) \times W) \\
&= \widetilde{\omega_3}^* \{\bar{c}\}^{k-h} \cdot ((Bq^* F^{h-1}) \times W) \ .
\end{aligned}$$

We have therefore shown the following proposition :

Proposition 1 *In* $CH^\bullet(\widehat{H^2(V)} \times V \times W)$, *one has the formula :*

$$\{\hat{u}_* s(\hat{U}) \times cW\}^m = \widetilde{\omega_3}^* \{\bar{c}\}^k \cdot (\hat{U} \times W) + \sum_{h \geq 1} (-2)^h \widetilde{\omega_3}^* \{\bar{c}\}^{k-h} \cdot ((Bq^* F^{h-1}) \times W) \ ,$$

where $\bar{c} = c(V)^{-1} \times cW$ *and* $k = m - n$.

3.2.5   Computation of $pr_{1*} \widetilde{\omega}_{1*} \nu_2$, first part

a) From notation 13, one has in $CH^\bullet(\widehat{H^2(V)} \times V \times W)$:

$$\nu_2 = \tilde{q}^* \alpha^* \widehat{H^2(\Gamma)} \cdot \{\hat{u}_* s(\hat{U}) \times cW\}^m \ .$$

But one knows that (see (2.2)) :

$$\alpha^* \widehat{H^2(\Gamma)} = \tilde{\pi}^* \widetilde{p_1}^* \Gamma \cdot (\tilde{\pi}^* \widetilde{p_2}^* \Gamma - \{s(F) \times cW\}^m) \ .$$

Moreover, for $i = 1, 2$, one has

$$\widetilde{\omega}_i = \tilde{p}_i \circ \tilde{\pi} \circ \tilde{q} \ . \tag{3.10}$$

Since $\tilde{q} = q \times id_W$, it follows that

$$\nu_2 = \widetilde{\omega}_1^* \Gamma \cdot (\widetilde{\omega}_2^* \Gamma - \{q^* s(F) \times cW\}^m) \cdot \{\hat{u}_* s(\hat{U}) \times cW\}^m \ .$$

Let us then decompose ν_2, by using the following notation :

Notation 15 : Let $\nu_2 = \nu_2' - \nu_2'' \in CH^\bullet(\widehat{H^2(V)} \times V \times W)$, where

$$\begin{aligned}
\nu_2' &= \widetilde{\omega}_1^* \Gamma \cdot \widetilde{\omega}_2^* \Gamma \cdot \{\hat{u}_* s(\hat{U}) \times cW\}^m \ . \\
\nu_2'' &= \widetilde{\omega}_1^* \Gamma \cdot \{q^* s(F) \times cW\}^m \cdot \{\hat{u}_* s(\hat{U}) \times cW\}^m \ .
\end{aligned}$$

In this paragraph, we are going to calculate $pr_{1*}\widetilde{\omega}_{1*}\nu'_2$.

From proposition 1, one has :

$$
\begin{aligned}
\nu'_2 =\ & \widetilde{\omega}_1^*\Gamma \cdot \widetilde{\omega}_2^*\Gamma \cdot \widetilde{\omega}_3^*\{\bar{c}\}^k \cdot (\widehat{U} \times W) \\
& + \sum_{h=1}^{k}(-2)^h \widetilde{\omega}_1^*\Gamma \cdot \widetilde{\omega}_2^*\Gamma \cdot \widetilde{\omega}_3^*\{\bar{c}\}^{k-h} \cdot ((B \cdot q^* F^{h-1}) \times W) ,
\end{aligned} \qquad (3.11)
$$

which we shorten as follows :

$$
\nu'_2 = \sum_{h=0}^{k} \nu'^{h}_2 . \qquad (3.12)
$$

b) Let us consider the first term in the sum (3.12) :

$$
\nu'^{0}_2 = \widetilde{\omega}_1^*\Gamma \cdot \widetilde{\omega}_2^*\Gamma \cdot \widetilde{\omega}_3^*\{\bar{c}\}^k \cdot (\widehat{U} \times W) .
$$

$\widehat{U}$ (see lemma 4.(ii)) is the union of G_{13} and G_{23}, the two graphs of the morphisms :

$$
\begin{aligned}
\widehat{H^2(V)} &\rightarrow V \\
(v_1, v_2, d) &\mapsto v_i \quad (i = 1, 2) .
\end{aligned}
$$

Let us use some additional notation :

Notation 16 : Let $\nu'^{0}_2 = a_1 + a_2$, where

$$
\begin{aligned}
a_1 &= \widetilde{\omega}_1^*\Gamma \cdot \widetilde{\omega}_2^*\Gamma \cdot \widetilde{\omega}_3^*\{\bar{c}\}^k \cdot (G_{13} \times W) \\
a_2 &= \widetilde{\omega}_1^*\Gamma \cdot \widetilde{\omega}_2^*\Gamma \cdot \widetilde{\omega}_3^*\{\bar{c}\}^k \cdot (G_{23} \times W) .
\end{aligned}
$$

Let us first study a_1 and $\widetilde{\omega}_{1*}a_1$.

Since on $(G_{13} \times W)$ the restrictions $\widetilde{\omega}_{1|}$ and $\widetilde{\omega}_{3|}$ are equal, one has :

$$
a_1 = \widetilde{\omega}_1^*(\Gamma\{\bar{c}\}^k) \cdot \widetilde{\omega}_2^*\Gamma \cdot (G_{13} \times W) .
$$

Applying (FP) to $\widetilde{\omega}_1$ yields :

$$
\widetilde{\omega}_{1*}a_1 = \Gamma\{\bar{c}\}^k \cdot \widetilde{\omega}_{1*}(\widetilde{\omega}_2^*\Gamma \cdot (G_{13} \times W)) .
$$

Since (see (3.10)) $\widetilde{\omega}_i = \tilde{p}_i \circ \tilde{\pi} \circ \tilde{q}$ for $i = 1, 2$, one has :

$$
\widetilde{\omega}_{1*}(\widetilde{\omega}_2^*\Gamma \cdot (G_{13} \times W)) = \widetilde{p}_{1*}\tilde{\pi}_*\tilde{q}_*(\tilde{q}^*\tilde{\pi}^*\widetilde{p}_2^*\Gamma \cdot (G_{13} \times W)) .
$$

But $\tilde{q}_*(G_{13} \times W) = [\widehat{H^2(V)} \times W] = \mathbf{1}$, since G_{13} is a <u>graph</u> of a morphism from $\widehat{H^2(V)}$ to V. Applying (FP) to $\tilde{q}$ yields :

$$\widetilde{\omega_1}_*(\widetilde{\omega_2}^*\Gamma \cdot (G_{13} \times W)) = \widetilde{p_1}_*\tilde{\pi}_*(\tilde{\pi}^*\widetilde{p_2}^*\Gamma \cdot \mathbf{1}) = \widetilde{p_1}_*\widetilde{p_2}^*\Gamma \cdot \mathbf{1}$$

(from (FP) and the fact that $\tilde{\pi}$ is birational), or $pr_2^* pr_{2*}\Gamma$ by lemma 2. But one has the equality $pr_{2*}\Gamma = f_*V$ (see (2.5)). By (3.7), it follows that :

$$\widetilde{\omega_1}_* a_1 = \Gamma \cdot \{\bar{c}\}^k \cdot pr_2^* f_*V = j_* j^*(\{\bar{c}\}^k \cdot pr_2^* f_*V) \quad .$$

It remains to compute

$$pr_{1*}\widetilde{\omega_1}_* a_1 = pr_{1*} j_*(j^*\{\bar{c}\}^k \cdot j^* pr_2^* f_*V) \quad .$$

Since $pr_{1*} j_* = \sigma_* = \gamma^*$, one has :

$$pr_{1*}\widetilde{\omega_1}_* a_1 = \gamma^* j^*\{\bar{c}\}^k \cdot \gamma^* j^* pr_2^* f_*V \quad .$$

On one hand, one has (2.8): $\gamma^* j^*\{\bar{c}\}^k = c_k$, on the other hand, $\gamma^* j^* pr_2^* = f^*$. The result

$$pr_{1*}\widetilde{\omega_1}_* a_1 = c_k f^* f_*V \tag{3.13}$$

follows.

Let us now study a_2 and $\widetilde{\omega_1}_* a_2$.

Since on $G_{23} \times W$, one has the equality $\widetilde{\omega_3}_| = \widetilde{\omega_2}_|$, one obtains :

$$a_2 = \widetilde{\omega_1}^*\Gamma \cdot \widetilde{\omega_2}^*(\Gamma\{\bar{c}\}^k) \cdot (G_{23} \times W)$$

Therefore one gets from (FP) :

$$\begin{aligned}
\widetilde{\omega_1}_* a_2 &= \Gamma \cdot \widetilde{\omega_1}_*(\widetilde{\omega_2}^*(\Gamma\{\bar{c}\}^k) \cdot (G_{23} \times W)) \\
&= \Gamma \cdot \widetilde{p_1}_*\tilde{\pi}_*\tilde{q}_*(\tilde{q}^*\tilde{\pi}^*\widetilde{p_2}^*(\Gamma\{\bar{c}\}^k) \cdot (G_{23} \times W)) \quad .
\end{aligned}$$

As was done above, we apply (FP) to $\tilde{q}$. Since G_{23} is the graph of a morphism from $\widehat{H^2(V)}$ to V, it follows that :

$$\tilde{q}_*[G_{23} \times W] = [\widehat{H^2(V)} \times W] = 1 \quad .$$

Next, we apply (FP) to $\tilde{\pi}$ which is birational. Hence :

$$\begin{aligned}
\widetilde{\omega_1}_* a_2 = \Gamma \cdot \widetilde{p_1}_*\widetilde{p_2}^*(\Gamma\{\bar{c}\}^k) &= \Gamma \cdot pr_2^* pr_{2*}(\Gamma\{\bar{c}\}^k) \quad \text{by lemma 2} \\
&= j_* j^* pr_2^* pr_{2*}(\Gamma\{\bar{c}\}^k) \quad \text{by (3.7)}.
\end{aligned}$$

Hence :

$$\begin{aligned}
pr_{1*}\widetilde{\omega_1}_* a_2 &= pr_{1*} j_* j^* pr_2^* pr_{2*}(\Gamma\{\bar{c}\}^k) \\
&= \gamma^* j^* pr_2^* pr_{2*}(j_* j^*\{\bar{c}\}^k), \quad \text{by (3.7) and since } pr_{1*} j_* = \sigma_* = \gamma^* \quad .
\end{aligned}$$

But one has (2.7): $\quad j^*\{\overline{c}\}^k = \sigma^* c_k = \gamma_* c_k$. The equalities

$$pr_{1*}\widetilde{\omega}_{1*} a_2 = \gamma^* j^* pr_2^* pr_{2*} j_* \gamma_* c_k = f^* f_* c_k$$

follow, since $f = pr_2 \circ j \circ \gamma$.

To summarize, since $\nu_2'^0 = a_1 + a_2$, we have obtained :

$$pr_{1*}\widetilde{\omega}_{1*}\nu_2'^0 = c_k f^* f_* V + f^* f_* c_k . \tag{3.14}$$

c) **Let us consider the other terms of the sum** (3.12):

We wrote above $\nu_2' = \sum_{h=0}^{k} \nu_2'^h$ with (for $h \geq 1$) $\nu_2'^h = (-2)^h w_h$, where

$$w_h = \widetilde{\omega}_1^* \Gamma \cdot \widetilde{\omega}_2^* \Gamma \cdot \widetilde{\omega}_3^* \{\overline{c}\}^{k-h} \cdot ((Bq^* F^{h-1}) \times W) .$$

Let $b : B \hookrightarrow \widehat{H^2(V)} \times V$ be the canonical imbedding ; then $Bq^* F^{h-1}$ can be rewritten as $b_* b^* q^* F^{h-1}$ from (FP). Since on B, the restrictions $\omega_{1|}$, $\omega_{2|}$ and $\omega_{3|}$ are equal, one can rewrite w_h as :

$$w_h = \widetilde{\omega}_1^* \Gamma \cdot \widetilde{\omega}_1^* \Gamma \cdot \widetilde{\omega}_1^* \{\overline{c}\}^{k-h} \cdot (b_* b^* q^* F^{h-1} \times W) .$$

It follows that in $CH^*(V \times W)$:

$$\widetilde{\omega}_{1*} w_h = \Gamma^2 \cdot \{\overline{c}\}^{k-h} \cdot \widetilde{\omega}_{1*}(b_* b^* q^* F^{h-1} \times W) .$$

But in $V \times W$, one has codim $\Gamma = m > n = $ dim Γ. Hence $\Gamma^2 = 0$. It follows that $\widetilde{\omega}_{1*} w_h = 0$. Therefore one has $\widetilde{\omega}_{1*}\nu_2'^h = 0$ for $h \geq 1$.

To conclude, we have obtained the result (see (3.14)) :

$$pr_{1*}\widetilde{\omega}_{1*}\nu_2' = c_k f^* f_* V + f^* f_* c_k . \tag{3.15}$$

3.2.6   Computation of $pr_{1*}\widetilde{\omega}_{1*}\nu_2$, second part

Notation 17 : Let $\delta : F \hookrightarrow \widehat{H^2(V)}$ be the canonical imbedding and let δ' be equal to $\delta \times id_V : F \times V \hookrightarrow \widehat{H^2(V)} \times V$.

a) From notation 15, one has $\quad \nu_2 = \nu_2' - \nu_2'' \quad$ where

$$\nu_2'' = \widetilde{\omega}_1^* \Gamma \cdot \{q^* \delta_* s(F) \times cW\}^m \cdot \{\hat{u}_* s(\widehat{U}) \times cW\}^m .$$

Recall that in chapter 1 we introduced the convention of omitting or not omitting the notation i_* if i is the canonical imbedding of a subscheme, depending on the case one considers.

In this paragraph, we calculate $pr_{1*}\widetilde{\omega}_{1*}\nu_2''$.

From proposition 1, one has in $CH^\bullet(\widehat{H^2(V)} \times V \times W)$:

$$\begin{aligned}
\nu_2'' &= \widetilde{\omega}_1^*\Gamma \cdot \{q^*\delta_*s(F) \times cW\}^m \cdot \widetilde{\omega}_3^*\{\overline{c}\}^k \cdot (\widehat{U} \times W) \\
&+ \sum_{h=1}^{k}(-2)^h\widetilde{\omega}_1^*\Gamma \cdot \{q^*\delta_*s(F) \times cW\}^m \cdot \widetilde{\omega}_3^*\{\overline{c}\}^{k-h} \cdot (Bq^*F^{h-1} \times W) \, . \quad (3.16)
\end{aligned}$$

One can rewrite ν_2'' as :

$$\nu_2'' = \sum_{h=0}^{k} \nu_2''^h \, . \tag{3.17}$$

b) **Let us consider the first term of the sum (3.17) :**

$$\nu_2''^0 = \widetilde{\omega}_1^*\Gamma \cdot \{q^*\delta_*s(F) \times cW\}^m \cdot \widetilde{\omega}_3^*\{\overline{c}\}^k \cdot (\widehat{U} \times W) \, .$$

Notice first (see notation 17) that $q^*\delta_*s(F) = \delta'_*(s(F) \times V)$. Also we have already noticed (lemma 4.(ii)) that $\widehat{U} = \cup_{i=1}^2 G_{i3}$. Since in $\widehat{H^2(V)} \times V \times W$ one has $\text{codim}(G_{i3} \times W) = n$, it follows that :

$$\begin{aligned}
\{q^*\delta_*s(F) \times cW\}^m \cdot (\widehat{U} \times W) &= \sum_{i=1}^{2}\{\delta'_*(s(F) \times V) \times cW\}^m \cdot (G_{i3} \times W) \\
&= \sum_{i=1}^{2}\{(\delta'_*(s(F) \times V) \cdot G_{i3}) \times cW\}^{m+n}, \quad \text{by } (\mathcal{R}_1).
\end{aligned}$$

From (FP), one has $\delta'_*(s(F) \times V) \cdot G_{i3} = \delta'_*((s(F) \times V) \cdot \delta'^*G_{i3})$. The computation performed in chapter 4 yields

$$\delta'^*G_{i3} = B \quad \text{for} \quad i = 1, 2 \, . \tag{3.18}$$

Hence :

$$\{q^*\delta_*s(F) \times cW\}^m \cdot (\widehat{U} \times W) = 2\{(\delta'_*(s(F) \times V) \cdot B) \times cW\}^{m+n} \, .$$

It follows that :

$$\nu_2''^0 = 2\widetilde{\omega}_1^*\Gamma \cdot \widetilde{\omega}_3^*\{\overline{c}\}^k \cdot \{(\delta'_*(s(F) \times V) \cdot B) \times cW\}^{m+n} \, . \tag{3.19}$$

As we have seen previously, on B, the restrictions $\omega_{1|}$, $\omega_{2|}$ and $\omega_{3|}$ are equal. Therefore, one has :

$$\nu_2''^0 = 2\widetilde{\omega}_1^*\Gamma \cdot \widetilde{\omega}_1^*\{\overline{c}\}^k \cdot \{(\delta'_*(s(F) \times V) \cdot B) \times cW\}^{m+n} \, .$$

Therefore, since $\text{codim}(\widetilde{\omega}_1) = -2n$ and $m - n = k$, (FP) and $(\mathcal{R}_2)$ yield :

$$\widetilde{\omega}_{1*}\nu_2''^0 = 2\Gamma \cdot \{\overline{c}\}^k \cdot \{\omega_{1*}(\delta'_*((s(F) \times V) \cdot B)) \times cW\}^k \, .$$

Let us consider the commutative diagram

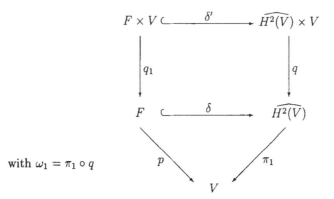

Diagram 5

where q_1 is the first projection (restriction of q) and p is the restriction of π_1 : $\widehat{H^2(V)} \to V$. Then one gets the equalities :

$$
\begin{aligned}
\omega_{1*}\delta'_*((s(F) \times V) \cdot B) &= p_*q_{1*}(q_1^*s(F) \cdot B) = p_*(s(F) \cdot q_{1*}B), \quad \text{by (FP)} \\
&= p_*(s(F) \cdot \mathbf{1}), \quad \text{because } B \text{ is a graph of } F \text{ in } V \\
&= c(V)^{-1}, \quad \text{because } F \text{ can be identified with the exceptional} \\
&\qquad \text{divisor of } \widetilde{V \times V}.
\end{aligned}
$$

Finally, since $\bar{c} = c(V)^{-1} \times cW$, we are left with :

$$
\widetilde{\omega_1}_* \nu_2''^0 = 2\Gamma \cdot \{\bar{c}\}^k \cdot \{\bar{c}\}^k = 2j_*j^*(\{\bar{c}\}^k)^2 \quad \text{by (3.7)} .
$$

Hence, as usual :

$$
pr_{1*}\widetilde{\omega_1}_* \nu_2''^0 = 2pr_{1*}j_*j^*(\{\bar{c}\}^k)^2 = 2\gamma^*j^*(\{\bar{c}\}^k)^2 \quad \text{car } pr_{1*}j_* = \sigma_* = \gamma^* .
$$

In view of (2.8), we have thus proved the result :

$$
pr_{1*}\widetilde{\omega_1}_* \nu_2''^0 = 2c_k^2 . \tag{3.20}
$$

c) **Let us look at the other terms of the sum (3.17) :**

Recall that $\nu_2'' = \sum_{h=0}^{k} \nu_2''^h$ where, for $h \geq 1$, one has

$$
\nu_2''^h = (-2)^h \widetilde{\omega_1}^* \Gamma \cdot \{q^*\delta_*s(F) \times cW\}^m \cdot \widetilde{\omega_3}^* \{\bar{c}\}^{k-h} \cdot (Bq^*F^{h-1} \times W) .
$$

But, as already seen in b), the restrictions $\omega_{1|B}$ and $\omega_{3|B}$ are equal. One can therefore substitute $\widetilde{\omega_3}^*$ by $\widetilde{\omega_1}^*$ in the above equation. Furthermore, since $\text{codim}(Bq^*F^{h-1} \times W) = n + h$, one has from $(\mathcal{R}_1)$:

$$
\begin{aligned}
\{q^*\delta_*s(F) \times cW\}^m \cdot (Bq^*F^{h-1} \times W) &= \{(q^*\delta_*s(F) \cdot Bq^*F^{h-1}) \times cW\}^{m+n+h} \\
&= \{(q^*(\delta_*s(F) \cdot F^{h-1}) \cdot B) \times cW\}^{m+n+h}
\end{aligned}
$$

Hence :

$$\nu_2''^h = (-2)^h \widetilde{\omega_1}^*(\Gamma\{\overline{c}\}^{k-h}) \cdot \{(q^*(\delta_*s(F) \cdot F^{h-1}) \cdot B) \times cW\}^{m+n+h} \quad .$$

Since $\mathrm{codim}(\widetilde{\omega_1}) = -2n$ and $k = m - n$, we obtain from (FP) and $(\mathcal{R}_2)$ the equality in $\mathrm{CH}^\bullet(V \times W)$:

$$\widetilde{\omega_1}_* \nu_2''^h = (-2)^h \Gamma\{\overline{c}\}^{k-h} \cdot \{\widetilde{\omega_1}_*(q^*(\delta_*s(F) \cdot F^{h-1}) \cdot B) \times cW\}^{k+h} \quad .$$

If one looks at the commutative diagram 5, since B is a graph from F to V, one see from (FP) that :

$$q_*(q^*(\delta_*s(F) \cdot F^{h-1}) \cdot B) = \delta_*s(F) \cdot F^{h-1} \cdot q_*B = \delta_*s(F) \cdot F^h \, , \quad \text{since} \quad q_*B = F \quad .$$

But (see [FU2], p. 70), one has :

$$\delta_*s(F) = \frac{F}{1+F} \quad \in \quad \mathrm{CH}^\bullet(\widehat{H^2(V)}) \quad . \tag{3.21}$$

Therefore one has :

$$\begin{aligned}(-1)^h q_*(q^*(\delta_*s(F) \cdot F^{h-1}) \cdot B) &= (-1)^h \delta_*s(F) \cdot F^h \\ &= (-1)^h \frac{F^{h+1}}{1+F} \\ &= \frac{F}{1+F} + (-F + (-F)^2 + \cdots + (-F)^h) \quad .\end{aligned}$$

Since $\omega_{1*} = \pi_{1*}q_*$, it yields :

$$\omega_{1*}((-1)^h(q^*(\delta_*s(F) \cdot F^{h-1}) \cdot B)) = \pi_{1*}\left(\frac{F}{1+F} + \sum_{i=1}^{h}(-F)^i\right) \quad .$$

But one has $\pi_{1*}F^i = 0$ for $1 \leq i \leq h$, since in $\widehat{H^2(V)}$,

$$\dim(F^i) = 2n - i \geq 2n - h \geq 2n - k > n$$

(Recall that $k \leq n/2$ from hypothesis 3). We have also seen (3.21) that :

$$\delta_*s(F) = \frac{F}{1+F} \qquad \text{hence} \qquad \pi_{1*}\left(\frac{F}{1+F}\right) = c(V)^{-1} \quad .$$

To summarize, one has :

$$\widetilde{\omega_1}_* \nu_2''^h = 2^h \Gamma \cdot \{\overline{c}\}^{k-h} \cdot \{c(V)^{-1} \times cW\}^{k+h} = 2^h j_* j^*(\{\overline{c}\}^{k-h} \cdot \{\overline{c}\}^{k+h}) \quad \text{by (3.7)}.$$

Hence, as usual :

$$pr_{1*}\widetilde{\omega_1}_* \nu_2''^h = 2^h pr_{1*}j_*(j^*\{\overline{c}\}^{k-h} \cdot j^*\{\overline{c}\}^{k+h}) = 2^h \gamma^* j^* \{\overline{c}\}^{k-h} \cdot \gamma^* j^* \{\overline{c}\}^{k+h}$$

because $pr_{1*}j_* = \sigma_* = \gamma^*$. (2.8) yields the result :

$$pr_{1*}\widetilde{\omega_1}_* \nu_2''^h = 2^h c_{k-h}c_{k+h} \quad . \tag{3.22}$$

In view of (3.20), we have finally found that :

$$pr_{1*}\widetilde{\omega_1}_* \nu_2'' = 2c_k^2 + \sum_{h=1}^{k} 2^h c_{k-h}c_{k+h} \quad . \tag{3.23}$$

3.2.7   Conclusion

We had from (3.6): $\widehat{m_3} = pr_{1*}\widetilde{\omega}_{1*}\nu$ and also, $\nu = \nu_1 - \nu_2 = \nu_1 - (\nu'_2 - \nu''_2)$ according to notation 13 and 15. Hence we have the equality $\widehat{m_3} = pr_{1*}\widetilde{\omega}_{1*}\nu_1 - pr_{1*}\widetilde{\omega}_{1*}\nu'_2 + pr_{1*}\widetilde{\omega}_{1*}\nu''_2$. Then, according to (3.8), (3.15) and (3.23), one has the equality :

$$\widehat{m_3} = m_2 f^* f_* V - c_k f^* f_* V - f^* f_* c_k + 2c_k^2 + \sum_{h=1}^{k} 2^h c_{k-h} c_{k+h} \quad . \tag{3.24}$$

Using the known expression for m_2 (see theorem 2), one can give for $\widehat{m_3}$ the more familiar expression (cf. Kleiman [KL1], th. 5.9) :

$$\widehat{m_3} = f^* f_* m_2 - 2c_k m_2 + \sum_{h=1}^{k} 2^h c_{k-h} c_{k+h} \quad .$$

To conclude, one has shown the following theorem :

Theorem 4 *Let $f : V \longrightarrow W$ be an arbitrary morphism of proper, smooth varieties with $\dim W = k + \dim V$ and $0 < k \leq \dim V /2$. Let $\widehat{m_3} \in CH^{2k}(V)$ be the "triple class", direct image of the cycle*

$$M_3 = [\widehat{H^3(\Gamma)}] \cdot [\widehat{H^3(V)} \times W] \quad \in \quad CH^\bullet(H^3(\widehat{V \times W}))$$

where Γ is the graph of f.

   Then one has the "triple formula"

$$\widehat{m_3} = f^* f_* m_2 - 2c_k m_2 + \sum_{h=1}^{k} 2^h c_{k-h} c_{k+h} \quad . \tag{3.25}$$

where c_i is the ith Chern class of the virtual normal bundle $f^* TW - TV$ and where m_2 denotes the double class in $CH^*(V)$.

Remark 5 : In chapter 5, we will explain (in a particular case) how to interpret the existence of excess components in the cycle M_3 in the case where the morphism f has S_2-singularities.

For example, let $f : \mathbb{P}^2 \longrightarrow \mathbb{P}^3$ be defined by $f[x : y : t] = [x^2 : xy : y^2 : t^2]$. Let H (resp. h) be the hyperplane generator of $CH^1(\mathbb{P}^3)$ (resp. $CH^1(\mathbb{P}^2)$). Then $f(\mathbb{P}^2)$ is a quadratic cone C and $f_*[\mathbb{P}^2] = 2[C] = 4H$. One has :

$$f^* H = 2h \quad \text{and} \quad c(f^* T\mathbb{P}^3 - T\mathbb{P}^2) = f^*(1 + H)^4 \cdot (1 + h)^{-3} = 1 + 5h + 6h^6 \quad .$$

The equality $m_2 = 3h$ follows (note that the double locus of f is degenerate here). Moreover, $f_* h = 2H^2$ and therefore (theorem 4) one has $\widehat{m_3} = 6h^2$, hence $\deg(\widehat{m_3}) = 6$. Here, the intersection is reduced to the only extra component I (defined in chapter 5, page 55), as is easily seen. This example is confirmed by theorem 5.

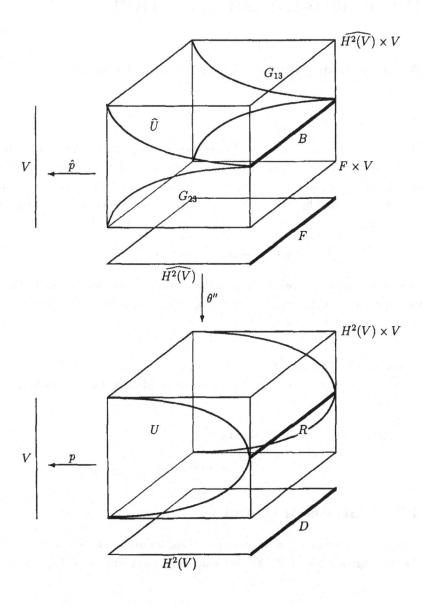

Chapter 4

Intermediate computations

This chapter provides details of the computations used previously.

4.1

Let X be a non-singular variety of dimension p and $d_0 \subset X$ a doublet of support a point $0 \in X$. One can choose local coordinates $(x, y_2, y_3, \ldots, y_p)$ centered at 0 such that the ideal of d_0 in $\mathcal{O}_X$ is $(x^2, y_2, y_3, \ldots, y_p)$. A chart of $H^2(X) = Hilb^2(X)$ at d_0 is then given (see [I1]) by $(a, b, c_2, d_2, \ldots, c_p, d_p)$, i.e. the coefficients of the neighbouring ideal in $\mathcal{O}_X$:

$$(x^2 + ax + b, \, -y_2 + c_2 x + d_2, \, -y_3 + c_3 x + d_3, \ldots, \, -y_p + c_p x + d_p) \quad .$$

(The minus signs are used to simplify the computations). Notice that in this chart, the hypersurface $D \subset H^2(X)$ consisting of doublets of support a single point is given by the equation $a^2 - 4b = 0$.

Notation 18 : One will write $\vec{y} = (y_2, \ldots, y_p)$, $\vec{c} = (c_2, \ldots, c_p)$ and $\vec{d} = (d_2, \ldots, d_p)$, so that the ideal of $d_0 \subset \mathcal{O}_X$ is $(x^2, \vec{y})$ and the ideal of a neighbouring doublet is $(x^2 + ax + b, \, -\vec{y} + x\vec{c} + \vec{d})$.

So a chart of $H^2(X)$ at d_0 is given by :

$$(a, \, b, \, \vec{c}, \, \vec{d}) \quad . \tag{4.1}$$

4.2 Flatness of π_1 and π_2

Let $d_0 \subset X$ be again a doublet of support 0 and let us consider $\widehat{H^2(X)} \subset H^2(X) \times X$, the tautological cover of $H^2(X)$. We want to give a chart of it at $\widehat{d_0} = (d_0, 0)$.

Let $(\xi, \eta_2, \ldots, \eta_p)$ be the coordinates of a point $p_1 \in X$ in the neighborhood of 0. Using the previous notation, one denotes again its coordinates by $(\xi, \vec{\eta})$. One expresses $\widehat{H^2(X)}$ in the chart $(a, b, \vec{c}, \vec{d})(\xi, \vec{\eta})$ of $H^2(X) \times X$ by requiring all coordinates of p_1 to verify the equations defining the doublet, i.e. :

$$\xi^2 + a\xi + b = 0 \tag{4.2}$$
$$-\vec{\eta} + \xi\vec{c} + \vec{d} = 0 \ . \tag{4.3}$$

One sees that $\widehat{H^2(X)}$ can be expressed locally as a graph $(\xi, \vec{\eta}, a, \vec{c}) \mapsto (b, \vec{d})$. And therefore

$$(\xi, \vec{\eta}, a, \vec{c}) \tag{4.4}$$

constitutes a chart of $\widehat{H^2(X)}$ at $(d_0, 0)$.

Let us then express the morphisms $\pi_1 : \widehat{H^2(X)} \longrightarrow X$ and $\pi_2 : \widehat{H^2(X)} \longrightarrow X$. For π_1, it is easy, since $\hat{d} = (d, p_1)$ must be sent to p_1. This gives :

$$\pi_1 : (\xi, \vec{\eta}, a, \vec{c}) \mapsto (\xi, \vec{\eta}) \ .$$

For π_2, one must find the coordinates of $p_2 = Res(p_1, d)$. The abscissa of p_2 is the other root of $x^2 + ax + b$, i.e. $-a - \xi$. As p_2 is located on the line defined by the equations $\vec{y} = x\vec{c} + \vec{d}$, its ordinate is $-(a + \xi)\vec{c} + \vec{d} = \vec{\eta} - (a + 2\xi)\vec{c}$, in view of (4.3). Hence :

$$\pi_1 : (\xi, \vec{\eta}, a, \vec{c}) \mapsto (\xi, \vec{\eta}) \tag{4.5}$$
$$\pi_2 : (\xi, \vec{\eta}, a, \vec{c}) \mapsto (-a - \xi, \vec{\eta} - (a + 2\xi)\vec{c}) \ . \tag{4.6}$$

These expressions show that π_1 and π_2 are submersions, and therefore flat morphisms.

The divisor $F \subset \widehat{H^2(X)}$ corresponding to the non simple doublets, i.e. $p_1 = p_2$, is then given by the equations :

$$\xi = -a - \xi \qquad \text{and} \qquad \vec{\eta} = \vec{\eta} - (a + 2\xi)\vec{c}$$

which reduce to

$$a + 2\xi = 0 \ . \tag{4.7}$$

4.3   Proof of lemma 4.(iv) and of $\Omega^1_{F|H} = 0$

How can the two-sheeted covering $\theta : \widehat{H^2(X)} \longrightarrow H^2(X)$, which is nothing else that the restriction to $\widehat{H^2(X)}$ of the natural projection $H^2(X) \times X \longrightarrow H^2(X)$, be expressed?

From equations (4.2) and (4.3), one has in the charts (4.4) and (4.1) :

$$\theta : (\xi, \vec{\eta}, a, \vec{c}) \mapsto (a, b = -\xi^2 - a\xi, \vec{c}, \vec{d} = \vec{\eta} - \xi\vec{c}) \ . \tag{4.8}$$

Notice in particular that if the ideal $(a^2 - 4b)$ of D is lifted by θ, one finds the ideal $(a + 2\xi)^2$, which is the square of the ideal of $F \subset \widehat{H^2(X)}$. Therefore, we have proved the statement (iv) of lemma 4 :

$$\theta^*[D] = 2[F] \quad \text{in} \quad \mathrm{CH}^1(\widehat{H^2(X)}) . \tag{4.9}$$

The Jacobian $D\theta$ of θ (in the charts (4.4) and (4.1)) is :

$$\begin{pmatrix} 1 & 0 & 0 & 0 \\ -\xi & -2\xi - a & 0 & 0 \\ 0 & 0 & \mathbf{I} & 0 \\ 0 & -\vec{c} & -\xi\mathbf{I} & \mathbf{I} \end{pmatrix}$$

where $\mathbf{I}$ denotes the identity matrix. $D\theta$ is invertible except if $2\xi + a = 0$, which is the equation of F; therefore we have proved (1.1) (see chapter 1). Moreover, the restriction of θ to F is not ramified, since in the local coordinates $(\xi, \vec{\eta}, \vec{c})$ of F, θ is given by :

$$(\xi, \vec{\eta}, \vec{c}) \mapsto (-2\xi, \xi^2, \vec{c}, \vec{\eta} - \xi\vec{c}) \quad .$$

This shows that :

$$\Omega^1_{F|H} = 0 , \tag{4.10}$$

which is used in the demonstration of lemma 5.

4.4   Proof of lemma 4.(iii)

Let us now prove some other statements of lemma 4 (by denoting again the variety by X and not by V). Let $(\xi', \vec{\eta}')$ be the coordinates of a point $p_3 \in X$.

In $\widehat{H^2(X)} \times X$, (4.4) provides a chart

$$(\xi, \vec{\eta}, a, \vec{c})(\xi', \vec{\eta}') \quad . \tag{4.11}$$

In this chart, the graph G_{13} of $\pi_1 : \widehat{H^2(X)} \longrightarrow X$ is given by the condition $p_3 = p_1$, i.e. :

$$\xi' = \xi \quad \text{and} \quad \vec{\eta}' = \vec{\eta} .$$

In the same chart, the graph G_{23} of π_2 (condition $p_3 = p_2$) is given by (see (4.6)) :

$$\xi' = -a - \xi \quad \text{and} \quad \vec{\eta}' = \vec{\eta} - (a + 2\xi)\vec{c} .$$

Consequently, one has the ideals :

$$I(G_{13}) = (\xi' - \xi, \vec{\eta}' - \vec{\eta}) \quad \text{and} \quad I(G_{23}) = (\xi' + a + \xi, \vec{\eta}' - \vec{\eta} + (a + 2\xi)\vec{c}) . \tag{4.12}$$

Therefore, the ideal of the scheme-theoretic intersection $G_{13} \cap G_{23}$ is :

$$(\xi' - \xi, \vec{\eta}' - \vec{\eta}, \xi' + a + \xi, \vec{\eta}' - \vec{\eta} + (a + 2\xi)\vec{c}) ,$$

or $(\xi' - \xi, \vec{\eta} - \vec{\eta}, a + 2\xi)$. This ideal corresponds to a non-singular subvariety of $\widehat{H^2(X)} \times X$. Of course, it is the same ideal as the ideal of the graph B of π_1 from F to X, since $B \subset G_{13} \cap G_{23}$ and B is non-singular. This proves the statement (iii) of lemma 4 :

$$B = G_{13} \cap G_{23} . \tag{4.13}$$

4.5   Proof of lemma 4.(ii) and (v)

Notice that the expression (4.8) locally gives

$$\theta' = \theta \times id_X : \qquad \widehat{H^2(X)} \times X \; \to \; H^2(X) \times X$$

$$(\xi, \vec{\eta}, a, \vec{c})(\xi', \vec{\eta}') \mapsto (a, b = -\xi^2 - a\xi, \vec{c}, \vec{d} = \vec{\eta} - \xi\vec{c})(\xi', \vec{\eta}') \tag{4.14}$$

a) Let us look for the inverse image $\widehat{U} = \theta'^{-1}(U)$ in $\widehat{H^2(X)} \times X$. In the chart $(a, b, \vec{c}, \vec{d})(\xi', \vec{\eta}')$ of $H^2(X) \times X$, the subvariety U is given by the ideal $(\xi'^2 + a\xi' + b, -\vec{\eta}' + \xi'\vec{c} + \vec{d})$, which can be obtained by replacing the coordinates $(\xi, \vec{\eta})$ by $(\xi', \vec{\eta}')$ in (4.2) and (4.3). By lifting it by θ', one finds the ideal of $\widehat{U}$:

$$((\xi' - \xi)(\xi' + \xi + a), \vec{\eta} - \vec{\eta}' + (\xi' - \xi)\vec{c}) .$$

Introduce the new coordinates $\mathcal{X} = \xi' - \xi$, $\mathcal{A} = \xi' + \xi + a$, $\vec{\mathcal{Y}} = \vec{\eta}' - \vec{\eta}$, keeping the other three ξ, $\vec{c}$ and $\vec{\eta}$. Therefore, the ideal of $\widehat{U}$ is

$$I(\widehat{U}) = (\mathcal{X}\mathcal{A}, -\vec{\mathcal{Y}} + \mathcal{X}\vec{c}) .$$

Also (see (4.12)), the ideal of $G_{13} \cup G_{23}$ in these new coordinates is :

$$I(G_{13} \cup G_{23}) = (\mathcal{X}, \vec{\mathcal{Y}}) \cap (\mathcal{A}, \vec{\mathcal{Y}} + (\mathcal{A} - \mathcal{X})\vec{c}) .$$

One does have $I(\widehat{U}) = I(G_{13} \cup G_{23})$, since the inclusion $\subset$ is obvious and the other one comes from the following computation : if $\alpha\mathcal{X} + \vec{\beta} \cdot \vec{\mathcal{Y}} = \gamma\mathcal{A} + \vec{\delta} \cdot (\vec{\mathcal{Y}} + (\mathcal{A} - \mathcal{X})\vec{c})$, where "$\cdot$" is the dot product, it follows that $\gamma\mathcal{A} + \mathcal{A}\vec{\delta} \cdot \vec{c} \in (\mathcal{X}, \vec{\mathcal{Y}})$.

Then $\gamma\mathcal{A} + \mathcal{A}\vec{\delta} \cdot \vec{c} = \lambda\mathcal{X} + \vec{\mu} \cdot \vec{\mathcal{Y}}$ and therefore λ (and $\vec{\mu}$) is a multiple of $\mathcal{A}$, by factoriality. Hence the condition $\gamma\mathcal{A} + \mathcal{A}\vec{\delta} \cdot \vec{c} \in (\mathcal{X}\mathcal{A}, -\vec{\mathcal{Y}} + \mathcal{X}\vec{c})$. Therefore, we have proved the equality of the schemes :

$$\widehat{U} = G_{13} \cup G_{23} . \tag{4.15}$$

b) Let us look for the inverse image $\theta'^*[R]$ where $R \subset U \subset H^2(X) \times X$ has been defined above (notation 12). First, one has the equations of U :

$$\xi'^2 + a\xi' + b = 0 \quad \text{and} \quad -\vec{\eta}' + \xi'\vec{c} + \vec{d} = 0 .$$

Furthermore, one has the equation of R itself, similar to (4.7): $a + 2\xi' = 0$. Thus the ideal of R is $(\xi'^2 + a\xi' + b, -\vec{\eta} + \xi'\vec{c} + \vec{d}, a + 2\xi')$. By lifting it by θ', one finds the ideal in $\widehat{H^2(X)} \times X$: $((\xi' - \xi)(\xi' + \xi + a), \vec{\eta} - \vec{\eta}' + (\xi' - \xi)\vec{c}, a + 2\xi')$.

Let us perform the same change of variables $\mathcal{X} = \xi' - \xi$, $\mathcal{A} = \xi' + \xi + a$, $\vec{\mathcal{Y}} = \vec{\eta}' - \vec{\eta}$, and keep ξ, $\vec{c}$ and $\vec{\eta}$. In these new coordinates, the ideal of $\theta'^{-1}(R)$ is

$$I = (\mathcal{X}\mathcal{A}, -\vec{\mathcal{Y}} + \mathcal{X}\vec{c}, \mathcal{A} + \mathcal{X}) \; .$$

On the other hand, the ideal of $B = G_{13} \cap G_{23}$ is, as was seen in § 4.4 :

$$J = (\xi' - \xi, \vec{\eta}' - \vec{\eta}, a + 2\xi) = (\mathcal{X}, \vec{\mathcal{Y}}, \mathcal{A} - \mathcal{X}) = (\mathcal{X}, \vec{\mathcal{Y}}, \mathcal{A}) \; .$$

In the new coordinates $(\mathcal{A}, \vec{c}, \xi, \vec{\eta}, \mathcal{X}, \vec{\mathcal{Y}})$, let M be the subvariety of ideal $K = (\vec{c}, \xi, \vec{\eta})$. Then one sees that $J + K$ is the ideal of the origin, while $I + K = (\mathcal{X}\mathcal{A}, \vec{\mathcal{Y}}, \mathcal{X} + \mathcal{A}, \vec{c}, \xi, \vec{\eta})$ is the ideal of a double point. It follows that :

$$\theta'^*[R] = 2[B] \; . \tag{4.16}$$

4.6   Proof of lemma 1

Now, we consider the situation of § 2.1, i.e. X is a subvariety of the non-singular variety Z. Let Π_1 and Π_2 : $\widehat{H^2(Z)} \longrightarrow Z$ be the two natural morphisms. If $F \subset \widehat{H^2(Z)}$ is the divisor of the non simple doublets, one writes :

$$\mathcal{D} = F \cap \Pi_1^{-1}(X) = F \cap \Pi_2^{-1}(X) \; .$$

Let us show that this intersection is transverse.

Let d_0 be a doublet of $\mathcal{D}$. We consider the most degenerate case : the case with the scheme theoretic inclusion of d_0 in X. If $p = \dim X$ and $q = \dim Z$, let $(x, y_2, \ldots, y_p, y'_{p+1}, \ldots, y'_q)$ be coordinates of Z (centered at $0 = supp(d_0)$), in which the equations of X are $y'_{p+1} = \cdots = y'_q = 0$ and the ideal of d_0 is $(x^2, y_2, \ldots, y_p, y'_{p+1}, \ldots, y'_q)$. We use a shortened notation, as in notation 18 :

$$(x, \vec{y}, \vec{y}') \quad \text{chart of } Z$$

where the ideal of d_0 is $I_0 = (x^2, \vec{y}, \vec{y}')$.

Let us use again the notation of § 4.2 and generalize it. As for (4.1), a chart of $H^2(Z)$ at d_0 is :

$$(a, b, \vec{c}, \vec{d}, \vec{c}', \vec{d}') \quad \text{where} \quad \vec{c}, \vec{d} \in \mathbb{C}^{p-1} \quad \text{and} \quad \vec{c}', \vec{d}' \in \mathbb{C}^{q-p} \tag{4.17}$$

correspond to the neighbouring ideal

$$I = (x^2 + ax + b, -\vec{y} + x\vec{c} + \vec{d}, -\vec{y}' + x\vec{c}' + \vec{d}') \; .$$

Let $(\xi, \vec{\eta}, \vec{\eta'})$ be the coordinates of a point of Z in the neighborhood of 0. Then one has the equations of $\widehat{H^2(Z)}$ in $H^2(Z) \times Z$:

$$\begin{cases} \xi^2 + a\xi + b = 0 \\ -\vec{\eta} + \xi\vec{c} + \vec{d} = 0 \\ -\vec{\eta'} + \xi\vec{c'} + \vec{d'} = 0 , \end{cases} \tag{4.18}$$

giving locally $\widehat{H^2(Z)}$ as a graph $(\xi, \vec{\eta}, \vec{\eta'}, a, \vec{c}, \vec{c'}) \mapsto (b, \vec{d}, \vec{d'})$. And therefore :

$$(\xi, \vec{\eta}, \vec{\eta'}, a, \vec{c}, \vec{c'}) \tag{4.19}$$

constitutes a chart of $\widehat{H^2(Z)}$ at $(d_0, 0)$. The morphism $\Pi_2 : \widehat{H^2(Z)} \longrightarrow Z$ can be expressed as (cf. (4.6)) :

$$(\xi, \vec{\eta}, \vec{\eta'}, a, \vec{c}, \vec{c'}) \mapsto (-a - \xi, \vec{\eta} - (a + 2\xi)\vec{c}, \vec{\eta'} - (a + 2\xi)\vec{c'}) .$$

Using the same argument as in § 4.2, one sees that in the chart (4.19), F is given by the equation $a + 2\xi = 0$; while $\Pi_2^{-1}(X)$ can be expressed by $\vec{\eta'} - (a + 2\xi)\vec{c'} = 0$.

Therefore, the intersection $\mathcal{D} = F \cap \Pi_2^{-1}(X) \subset \widehat{H^2(Z)}$ is transverse, $\tag{4.20}$

which proves lemma 1, page 14.

4.7 Flatness of P_{12} and of P_3

From now on, we concentrate on the variety $\widehat{H^3(X)}$ where X is non-singular (see § 3.1). It consists of the elements $\hat{t} = (t, d_{12}, d_{23}, d_{31}, p_1, p_2, p_3)$. Let

$$\begin{array}{ccccccc} P_{12} : & \widehat{H^3(X)} & \rightarrow & \widehat{H^2(X)} & \text{and} & P_3 : & \widehat{H^3(X)} & \rightarrow & X \\ & \hat{t} & \mapsto & \widehat{d_{12}} & & & \hat{t} & \mapsto & p_3 \end{array}$$

be the natural morphisms, where $\widehat{d_{12}} = (p_1, d_{12})$. We show that they are flat.

We are led to consider the neighborhood of the most degenerate case, i.e.

$$\hat{t}_o = (t_o, d_o, d_o, d_o, 0, 0, 0)$$

where t_o is an amorphous triplet. Recall the following definition (already given in the introduction – see 0.7–) :

A triplet $t \subset X$ is said to be amorphous if supp(t) is reduced to only one point p and if the ideal of t is the square of the ideal of p in a germ of a smooth surface going through p.

(The other triple points are curvilinear, i.e. subschemes of a non-singular curve). Consequently, in a chart $(x, y, \vec{z})$ of X (where $\vec{z} \in \mathbb{C}^{p-2}$ if $p = \dim X$), the ideal of the triplet t_o is $(x^2, xy, y^2, \vec{z})$ and the ideal of the doublet d_o is $(x^2, y, \vec{z})$. From ([LB1], p. 937), a chart of $\widehat{H^3(X)}$ at $\hat{t_o}$ is

$$(s, t, \vec{r}, c, c', c'', v, \vec{\rho}, \vec{\sigma}) . \tag{4.21}$$

Notice that in [LB1] it was assumed that $p = \dim X = 3$, but the generalization can be easily performed.

However, a chart of $H^2(X)$ at d_o is $(a, b, c, d, \vec{e}, \vec{f})$, which corresponds to the ideal $(x^2 + ax + b, -y + cx + d, -\vec{z} + \vec{e}x + \vec{f})$. (It is similar to notation 18, except that not only x but both x and y are needed in the calculation ; this is why $\vec{z}$ is introduced and not $\vec{y}$).

Let $(\xi, \eta, \vec{\zeta})$ be the coordinates of a point close to 0 ; thus a chart of $\widehat{H^2(X)}$ at $\widehat{d_o} = (d_o, 0)$ is :

$$(\xi, \eta, \vec{\zeta}, a, c, \vec{e}) \quad , \tag{4.22}$$

which comes from the equations of $\widehat{H^2(X)}$ in $H^2(X) \times X$:

$$\xi^2 + a\xi + b = 0 , \qquad -\eta + c\xi + d = 0 , \qquad -\vec{\zeta} + \vec{e}\xi + \vec{f} = 0$$

which state that the point must be on the doublet. (The calculation is exactly the same as for (4.2), (4.3) and (4.4)).

Then one gives the local expression of P_{12}. First, p_1 is the point of coordinates $(s, t, \vec{r})$. From [LB1], pp 934 and 937, one also has :

$$\begin{cases} a = -2s - s' = -2s + c''v \\ \vec{e} = -\vec{\rho} - \vec{\sigma}c . \end{cases}$$

The local expression of P_{12} in the charts (4.21) and (4.22) follows :

$$\hat{t} = (s, t, \vec{r}, c, c', c'', v, \vec{\rho}, \vec{\sigma}) \mapsto \widehat{d_{12}} = (s, t, \vec{r}, -2s + c''v, c, -\vec{\rho} - \vec{\sigma}c)$$

which proves, in the neighborhood of $\hat{t_o}$:

$$\text{the flatness of } P_{12}. \tag{4.23}$$

Notice once more that we studied P_{12} in the neighborhood of the most degenerate case ; in the other cases, the computations are similar.

The expression for P_{12} which was obtained previously enables us to show that :

$$P_{12}^{-1}(F) = E_{12} + E^\bullet \subset \widehat{H^3(X)} \quad \text{(scheme theoretically)} \tag{4.24}$$

Indeed, the divisor $F \subset \widehat{H^2(X)}$ of the non simple doublets is given locally in the chart (4.22) by $a + 2\xi = 0$ (same calculation as for (4.7)). If this equation is lifted by P_{12}, one finds $-2s + c''v + 2s = 0$, i.e. $c''v = 0$. From ([LB1], p. 940), it is indeed the equation of $E_{12} + E^\bullet$.

For P_3, it is in fact a submersion, since the coordinates of the point p_3 are :

$$(s + s' + s'', t + t' + t'', \vec{r} + \vec{r'} + \vec{r''})$$

(see [LB1], p. 935). Then one sees (again from [LB1], (1), (2), (13), (14) and (E)), that $p_3 = (s + \cdots, t + \cdots, \vec{r} + \cdots)$, where the dots denote terms of degree ≥ 2 in the coefficients of the chart (4.21).

$$\text{This proves the flatness of } P_3. \tag{4.25}$$

4.8   Proof of lemma 4.(i)

Now, let $\phi = P_{12} \times P_3 : \widehat{H^3(X)} \longrightarrow \widehat{H^2(X)} \times X$ be the morphism defined by $\phi(\hat{t}) = (\widehat{d_{12}}, p_3)$ and let $\widehat{U} \subset \widehat{H^2(X)} \times X$ be the inverse image of the tautological scheme $U \subset H^2(X) \times X$ by $\theta' = \theta \times id_X$ (see notation 12). We are going to show that scheme-theoretically, $\phi^{-1}(\widehat{U}) = \overline{E} = E_{13} + E_{23} + E^{\bullet}$. Once again, we just study the neighborhood of the most degenerate case : $\hat{t}_o = (t_o, d_o, d_o, d_o, 0, 0, 0)$, with t_o amorphous (cf. definition 2).

One uses again the previous notation : $(a, b, c, d, \vec{e}, \vec{f})$ is a chart of $H^2(X)$ at d_o ; let $(\xi', \eta', \vec{\zeta'})$ be the coordinates of a point of X close to 0, so that one has :

$$(\xi', \eta', \vec{\zeta'}) \qquad \text{chart or } X \text{ at } 0. \tag{4.26}$$

The ideal of U in the corresponding chart of $H^2(X) \times X$ is

$$(\xi'^2 + a\xi' + b, -\eta' + c\xi' + d, -\vec{\zeta'} + \vec{e}\xi' + \vec{f}) .$$

Moreover, the local expression of $\theta' = \theta \times id_X$ (similar to (4.14)) is

$$(\xi, \eta, \vec{\zeta}, a, c, \vec{e})(\xi', \eta', \vec{\zeta'}) \mapsto (a, -\xi^2 - a\xi, c, \eta - \xi c, \vec{e}, \vec{\zeta} - \vec{e}\xi)(\xi', \eta', \vec{\zeta'}) .$$

By lifting the ideal of U by θ', one finds (similarly to § 4.5) the ideal of $\widehat{U}$ in $\widehat{H^2(X)} \times X$:

$$((\xi' - \xi)(\xi' + \xi + a), (\xi' - \xi)c + \eta - \eta', (\xi' - \xi)\vec{e} + \vec{\zeta} - \vec{\zeta'}) .$$

The local expression of $\phi : \widehat{H^3(X)} \longrightarrow \widehat{H^2(X)} \times X$ in the charts (4.21) and (4.22)$\times$(4.26)

$$\phi : (s, t, \vec{r}, c, c', c'', v, \vec{p}, \vec{\sigma}) \mapsto (s, t, \vec{r}, -2s + c''v, c, -\vec{p} - \vec{\sigma}c)(s + s' + s'', t + t' + t'', \vec{r} + \vec{r'} + \vec{r''})$$

is now available (see § 4.7), where $s', s'', t', t'', \vec{r'}, \vec{r''}$ are given by ([LB1], relations (1), (2), (13), (14) and (E)) :

$$\begin{cases} s' &= -c''v \\ s'' &= -c'v \\ t' &= cs' \\ t'' &= s''(c + c' + c'') \end{cases} \qquad \begin{cases} \vec{r'} &= \vec{e}s' \\ \vec{r''} &= s''(\vec{e} + \vec{e'} + \vec{e''}) \\ \vec{e} &= -\vec{\rho} - \vec{\sigma}c \\ \vec{e'} &= -\vec{\sigma}c' \\ \vec{e''} &= -\vec{\sigma}c'' \ . \end{cases} \tag{4.27}$$

Then by lifting the ideal of $\widehat{U}$ by ϕ, one finds the ideal of $\phi^{-1}(\widehat{U}) \subset \widehat{H^3(X)}$:

$$((s' + s'')(s' + s'' + c''v), \ (s' + s'')c - t' - t'', \ (s' + s'')(\vec{\rho} + \vec{\sigma}c) + \vec{r'} + \vec{r''}) \ ,$$

i.e., by using the relations (4.27) :

$$((c' + c'')c'v^2, \ (c' + c'')c'v, \ (c' + c'')c'v\vec{\sigma}) = ((c' + c'')c'v) \ .$$

From [LB1], pp 939-940, one recognizes the equation of the divisor $\overline{E} = E_{13} + E_{23} + E^\bullet$. Hence :

$$\phi^{-1}(\widehat{U}) = \overline{E} \subset \widehat{H^3(X)} \ , \tag{4.28}$$

which is used in lemma 4.

4.9   Proof of lemma 3

We now consider the situation of § 4.6 : the variety X is now a subvariety (of dimension p) of the non-singular variety Z (of dimension q). The two natural morphisms are denoted once again by P_{12} and P_3 :

$$P_{12} : \widehat{H^3(Z)} \longrightarrow \widehat{H^2(Z)} \qquad \text{and} \qquad P_3 : \widehat{H^3(Z)} \longrightarrow Z \ .$$

Then, let $(x, y, \vec{z}, \vec{z'})$ be a chart of Z at a point, so that   (4.29)

$\vec{z'} = 0$ are equations of X in Z.

Let $\hat{t}_o = (t_o, d_o, d_o, d_o, 0, 0, 0)$ be a complete triple in X, with t_o amorphous ; it is a fortiori in Z. Let $\widehat{d}_o = (d_o, 0) \in \widehat{H^2(X)}$.

Similarly to (4.22), one obtains a chart of $\widehat{H^3(Z)}$ at $\hat{t}_o$, simply by changing $\vec{z}$ into $(\vec{z}, \vec{z'})$, therefore by changing $\vec{r}$ into $(\vec{r}, \vec{r'})$, $\vec{\rho}$ into $(\vec{\rho}, \vec{\rho'})$ and $\vec{\sigma}$ into $(\vec{\sigma}, \vec{\sigma'})$. The chart of $\widehat{H^3(Z)}$ at $\hat{t}_o$ follows :

$$(s, t, \vec{r}, \vec{r'}, c, c', c'', v, \vec{\rho}, \vec{\rho'}, \vec{\sigma}, \vec{\sigma'}) \ .$$

(**Warning** : the variable $\vec{r'}$ has nothing to do with $\vec{r'}$ used in the previous section !). How is $\widehat{H^3(X)}$ expressed in this chart ? Simply by $\vec{r'} = 0, \vec{\rho'} = 0, \vec{\sigma'} = 0$ and therefore :

$$\widehat{H^3(X)} \text{ is a non-singular subvariety of } \widehat{H^3(Z)}. \tag{4.30}$$

It remains to check that $P_{12}^{-1}(\widehat{H^2(X)})$ intersects transversally the divisors E_{13}, E_{23} and $E^{\bullet}$ in $\widehat{H^3(Z)}$. Of course, one uses the chart of $\widehat{H^2(Z)}$ at $\widehat{d_o}$, similar to (4.22) :

$$(\xi,\, \eta,\, \vec{\zeta},\, \vec{\zeta'},\, a,\, c,\, \vec{e},\, \vec{e'})\ .$$

In this chart, $\widehat{H^2(X)}$ is given by $\vec{\zeta'} = 0$, $\vec{e'} = 0$.

Then one uses the local expression, similar to (4.23) of P_{12} :

P_{12} :

$$\widehat{H^3(Z)} \qquad\qquad \to \qquad \widehat{H^2(Z)}$$
$$(s,t,\vec{r},\vec{r'},c,c',c'',v,\vec{\rho},\vec{\rho'},\vec{\sigma},\vec{\sigma'}) \mapsto (s,t,\vec{r},\vec{r'},-2s+c''v,c,-\vec{\rho}-\vec{\sigma}c,-\vec{\rho'}-\vec{\sigma'}c)$$

Thus $P_{12}^{-1}(\widehat{H^2(X)})$ has for equations $\vec{r'} = 0$ and $\vec{\rho'} + \vec{\sigma'}c = 0$ in $\widehat{H^3(Z)}$. One has :

$$E_{13},\ E_{23}\ \text{and}\ E^{\bullet}\ \text{are transverse at}\ P_{12}^{-1}(\widehat{H^2(X)})\ \text{in}\ \widehat{H^3(Z)}, \qquad (4.31)$$

since their respective equations are $c' + c'' = 0$, $c' = 0$ and $v = 0$.

4.10   Transversality of β and $\overline{\mathbb{E}}$

a) Finally, let V and W be two non-singular varieties, of respective dimensions n and m.

Let $\mathbb{F} \subset H^2(\widehat{V \times W})$ be the divisor of the non simple doublets. Let us prove that $\alpha^*\mathbb{F} = F \times W$, if $\alpha : \widehat{H^2(V)} \times W \hookrightarrow H^2(\widehat{V \times W})$ is the imbedding onto the horizontal doublets (see § 2.2).

Let $w \in W$ and $\widehat{d_o} = (d_o, 0) \in \widehat{H^2(V)}$, with d_o of support 0. Consider $\widehat{d'_o} = \alpha(\widehat{d_o}, w)$. At $(0, w)$, one has a chart of $V \times W : (x, \vec{y}, \vec{y'})$ where $(x, \vec{y})$ is a chart of V at 0 and $\vec{y'}$ is a chart of W at w, the ideal of $d_o \subset \mathcal{O}_V$ being $(x^2, \vec{y})$. In exactly the same way as in § 4.6, a chart of $H^2(\widehat{V \times W})$ at $\widehat{d'_o}$ is $(\xi,\, \eta,\, \vec{\eta'},\, a,\, \vec{c},\, \vec{c'})$, which corresponds to the doublets close to (d_o, w), of ideal

$$(x^2 + ax + b,\ -\vec{y} + x\vec{c} + \vec{d},\ -\vec{y'} + x\vec{c'} + \vec{d'})\ .$$

In this chart, $\mathbb{F}$ is given, as usual, by $a + 2\xi = 0$. However, $\vec{c'} = 0$ is the equation of $\alpha(\widehat{H^2(V)} \times W)$, since it must correspond to the horizontal doublets, and then $\vec{y'}$ is constant. This proves the statement :

$$\text{The intersection}\ \mathbb{F} \cap (\widehat{H^2(V)} \times W)\ \text{is transverse in}\ H^2(\widehat{V \times W}). \qquad (4.32)$$

b) Now let $\beta : \widehat{H^3(V)} \times W \hookrightarrow H^3(\widehat{V \times W})$ be the imbedding on the horizontal triplets. One would like to show that it is transverse to $\overline{\mathbb{E}} = \mathbb{E}_{13} + \mathbb{E}_{23} + \mathbb{E}^{\bullet}$.

Once again, we study what happens close to the most degenerate case : $\widehat{t_o} = (t_o, d_o, d_o, d_o, 0, 0, 0)$ where t_o is amorphous. Let $w \in W$. Consider $\widehat{t'_o} = \beta(\widehat{t_o}, w)$. At

$(0, w)$, one has a chart of $V \times W$: $(x, y, \vec{z}, \vec{z'})$, where $(x, y, \vec{z})$ is a chart of V at 0 and $\vec{z'}$ is a chart of W at w, the ideal of t_o being $(x^2, xy, y^2, \vec{z})$ and the ideal of d_o being $(x^2, y, \vec{z})$.

Similarly to (4.21), one obtains a chart of $H^3(\widehat{V \times W})$ at $\widehat{t'_o}$:

$$(s, t, \vec{r}, \vec{r'}, c, c', c'', v, \vec{\rho}, \vec{\rho'}, \vec{\sigma}, \vec{\sigma'}) \ .$$

In this chart, $\widehat{H^3(V)} \times W$ is expressed by $\vec{\rho'} = 0$ and $\vec{\sigma'} = 0$, since $\vec{z'}$ must be constant. The equations of the divisors $\mathbb{E}_{13}$, $\mathbb{E}_{23}$ and $\mathbb{E}^\bullet$ are respectively : $c' + c'' = 0$, $c' = 0$ and $v = 0$. It follows that

β is transverse to $\mathbb{E}_{13}$, $\mathbb{E}_{23}$ and $\mathbb{E}^\bullet$ and then $\beta^*\mathbb{E} = \overline{E} \times W$ in $\widehat{H^3(V)} \times W.$ (4.33)

Chapter 5

Application to the case where V is a surface and W a volume

Let V be a surface and W be a volume (i.e. $\dim V = 2$ and $\dim W = 3$) and let $f : V \longrightarrow W$ be a morphism having a S_2-singularity at the point $0 \in V$ which means that in local coordinates, f is written as :

$$f(x, y) = (q_1(x, y) + \cdots, q_2(x, y) + \cdots, q_3(x, y) + \cdots)$$

where the q_i are quadratic forms, the dots denoting terms of degree > 2.

The intersection $\widehat{H^3(\Gamma)} \cap (\widehat{H^3(V)} \times W)$ in $H^3(\widehat{V \times W})$ then possesses an excess component I, which consists of the

$$\widehat{T} = (T, D_{12}, D_{23}, D_{31}, p_1, p_2, p_3)$$

with $p_1 = p_2 = p_3 = 0$. The ideal of T is $\mathcal{M}^2$, where $\mathcal{M}$ is the ideal of 0 in $\mathcal{O}_V$. On the other hand, D_{12}, D_{23} and D_{31} are arbitrary and therefore set theoretically, one has a bijection :

$$\begin{aligned} I &\xrightarrow{\sim} \mathbb{P}^1 \times \mathbb{P}^1 \times \mathbb{P}^1 \\ \widehat{T} &\mapsto (D_{12}, D_{23}, D_{31}) , \end{aligned} \tag{5.1}$$

by writing $\mathbb{P}^1$ for $\mathbb{P}(T_0 V)$, where $\mathbb{P}$ denotes the set of lines.

In the following, we assume that I is reduced, which implies that (5.1) is an isomorphism. One has the commutative diagram where the arrows are canonical imbeddings (the dimensions are shown in parentheses) :

$$(9) \quad \widehat{H^3(V)} \times W \lhook\joinrel\longrightarrow H^3(\widehat{V \times W}) \quad (15)$$

$$\Big\downarrow v \qquad\qquad\qquad\qquad \Big\downarrow w$$

$$(3) \qquad\qquad I \quad\lhook\joinrel\longrightarrow\quad \widehat{H^3(\Gamma)} \qquad (6)$$

Then, the corollary 9.2.3 of [FU2], p. 163 yields :

$$[\widehat{H^3(\Gamma)}] \cdot [\widehat{H^3(V)} \times W] = c_3(\nu(w)_{|I} - \nu(v)) + \mathbb{R}$$

where ν denotes the normal bundle and $\mathbb{R}$ the "*residual class*".

5.1 Computation of $c\nu(v)$

a) Let $0' = f(0) \in W$. First, I is contained inside $\widehat{H^3(V)}$ (identified with $\widehat{H^3(V)} \times \{0'\}$) and even contained inside $E^\bullet$, so that v can be decomposed in the canonical imbeddings :

$$I \hookrightarrow E^\bullet \hookrightarrow \widehat{H^3(V)} \hookrightarrow \widehat{H^3(V)} \times W .$$

Thus, in the Grothendieck group $K(I)$ of the vector bundles, one has

$$\nu(v) = \nu(I, E^\bullet) + \nu(E^\bullet, \widehat{H^3(V)})_{|I} + \text{ trivial bundle.}$$

But $E^\bullet$ is a fibration on V (by $\widehat{T} \mapsto p_1$) and I is the fiber at 0. Thus, in fact,

$$\nu(v) = \nu(E^\bullet, \widehat{H^3(V)})_{|I} + \text{ trivial bundle },$$

which gives the total Chern class :

$$c\nu(v) = c\nu(E^\bullet, \widehat{H^3(V)})_{|I} \quad .$$

If $i : E^\bullet \hookrightarrow \widehat{H^3(V)}$ is the canonical imbedding, one has therefore

$$c_1\nu(E^\bullet, \widehat{H^3(V)}) = i^*[E^\bullet] . \tag{5.2}$$

Hence $c\nu(E^\bullet, \widehat{H^3(V)}) = 1 + i^*[E^\bullet]$. Abusing the notation, one has therefore :

$$c\nu(v) = 1 + [E^\bullet]_{|I} \quad . \tag{5.3}$$

Thus, it remains to find $[E^\bullet]_{|I}$ in $\mathrm{CH}^1(I)$.

Notation 19 : In $\mathrm{CH}^1(\mathbb{P}^1 \times \mathbb{P}^1 \times \mathbb{P}^1)$, one writes $A = pr_1^*(\bullet)$, $B = pr_2^*(\bullet)$, $C = pr_3^*(\bullet)$, so that :

$$\mathrm{CH}^1(\mathbb{P}^1 \times \mathbb{P}^1 \times \mathbb{P}^1) = \mathbb{Z}A \oplus \mathbb{Z}B \oplus \mathbb{Z}C \quad .$$

Notice that

$$A^2 = B^2 = C^2 = 0 \quad \text{and} \quad \deg(ABC) = 1 . \tag{5.4}$$

For symmetry reasons, one has

$$[E^\bullet]_{|I} = \alpha(A + B + C) \quad \text{with} \quad \alpha \in \mathbb{Z} . \tag{5.5}$$

Let $\delta \subset I \simeq \mathbb{P}^1 \times \mathbb{P}^1 \times \mathbb{P}^1$ be the "small" diagonal consisting of complete triples $(0^2, D, D, D, 0, 0, 0)$ where 0^2 denotes (improperly) the triplet of ideal $\mathcal{M}^2$. Obviously, one has $\delta \simeq \mathbb{P}^1$.

$b)$

Lemma 6 *By identifying δ with $\mathbb{P}^1$, one has the equality $c_1\nu(\delta, \widehat{H^3(V)}) = \mathbf{1}$.*

Proof :

Let (x, y) be the coordinates of V centered at 0. If D is a doublet of support 0, one denotes by $\mathrm{Axis}(D)$ the line it defines in this coordinate system. One sees that $\delta \simeq \mathbb{P}^1$ is the glueing of two open sets U_0 and U_∞ (each one is isomorphic to $\mathbb{C}$), where :

U_0 corresponds to the doublets D of non vertical axis,

U_∞ corresponds to the doublets D of non horizontal axis.

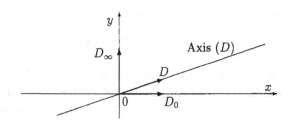

In [LB1], p. 937 was given a chart of $\widehat{H^3(V)}$ at $\widehat{T_0} = (0^2, D_0, D_0, D_0, 0, 0, 0)$ where D_0 is the doublet of ideal (x^2, y). This chart is

$$(s, t, c, c', c'', v) \quad , \tag{5.6}$$

with the notation of [LB1]. More precisely :

(i) (s, t) are the coordinates in the chart (x, y) of the point p_1 close to 0,

(ii) One denotes by

c	the slope of $\mathrm{Axis}(D_{12})$, of equation $y = cx + d$
$c + c'$	the slope of $\mathrm{Axis}(D_{31})$, of equation $y = (c + c')x + d + d'$
$c + c' + c''$	the slope of $\mathrm{Axis}(D_{23})$, of equation $y = (c + c' + c'')x + d + d' + d''$.

(*iii*) Finally v is a coefficient which arises in the ideal J of a triplet close to 0^2 :

$$J = (x^2 + ux + vy + w, \; xy + u'x + v'y + w', \; y^2 + u''x + v''y + w'') \quad .$$

Then let D_∞ be the doublet of ideal (y^2, x). If one considers the complete triple $\widehat{T_\infty} = (0^2, D_\infty, D_\infty, D_\infty, 0, 0, 0)$, one sees that a chart of $\widehat{H^3(V)}$ at $\widehat{T_\infty}$ is given by

$$(S, T, C, C', C'', V) \quad \text{where} \tag{5.7}$$

(*i*)' $S = t$, $T = s$ are the coordinates of the point p_1 in the chart (y, x) of V (and not (x, y)) ;

(*ii*)' moreover $C = \dfrac{1}{c}$ since in the chart (y, x), the equation of the line $\text{Axis}(D_{12})$ is

$$x = \frac{y}{c} - \frac{d}{c} \quad \text{(from (\textit{ii}))}. \text{ Also, from (\textit{ii}) again, } C + C' = \frac{1}{c + c'} \text{ and } C + C' + C'' = \frac{\frac{1}{c}}{c + c' + c''}.$$

(*iii*)' Finally, one has $u'' = V$, since the ideal J can be rewritten, by exchanging the roles of x and y, as

$$J = (y^2 + v''y + u''x + w'', \; yx + v'y + u'x + w', \; x^2 + vy + ux + w) \; .$$

The first generator of J should indeed be written as $y^2 + Uy + Vx + W$. But, from [LB1], relations (E), p. 937, one has :

$$u'' = -cv(c + c')(c + c' + c'') \; . \tag{5.8}$$

In the chart (5.6), δ is given by the equations $s = t = c' = c'' = v = 0$, i.e. δ is parameterized by the direction c of the line $y = cx$. On the other hand, in the chart (5.7), δ is given by the equations $S = T = C' = C'' = V = 0$. Thus, one obtains the normal bundle to δ in $\widehat{H^3(V)}$ as the glueing of $\mathbb{C}^* \times \mathbb{C}^5$ and $\mathbb{C}^* \times \mathbb{C}^5$ by :

$$(c, \, ds, \, dt, \, dc', \, dc'', \, dv) \mapsto (C, \, dS, \, dT, \, dC', \, dC'', \, dV) \; .$$

But $C = \dfrac{1}{c}$ and also, from (*ii*)' : $dC + dC' = \dfrac{-dc - dc'}{c^2}$, hence $dC' = \dfrac{-dc'}{c^2}$. Similarly, one has the equality $dC'' = \dfrac{-dc''}{c^2}$. Then, from (5.8), $dV = du'' = -c^3 dv$. Finally, from (*i*)', one has : $dS = dt$ and $dT = ds$. One obtains the glueing data :

		ds	dt	dc'	dc''	dv
	dS	0	1	0	0	0
	dT	1	0	0	0	0
$c \mapsto$	dC'	0	0	$-1/c^2$	0	0
	dC''	0	0	0	$-1/c^2$	0
	dV	0	0	0	0	$-c^3$

$$\mathbb{C}^* \; \rightarrow \qquad\qquad GL(5, \mathbb{C})$$

which gives $\nu(\delta, \widehat{H^3(V)})$ as bundle of rank 5 on $\mathbb{P}^1$.

But recall that the glueing of $\mathbb{C}^* \times \mathbb{C}$ with $\mathbb{C}^* \times \mathbb{C}$ by $(z, \xi) \mapsto (\frac{1}{z}, \frac{\xi}{z^n})$, where $n \in \mathbb{Z}$, gives a vector bundle on $\mathbb{P}^1$, whose first Chern class is n : section 1 has a <u>zero</u> of order n at infinity. Here, one sees that the first Chern class of $\nu(\delta, \widehat{H^3(V)})$ is $2 + 2 - 3 = 1$. $\qquad\square$

c) The inclusions $\delta \subset E^\bullet \subset \widehat{H^3(V)}$ yield in the Grothendieck group $K(\delta)$:

$$\nu(\delta, E^\bullet) + \nu(E^\bullet, \widehat{H^3(V)})_{|\delta} = \nu(\delta, \widehat{H^3(V)}) \quad .$$

Considering the first Chern classes and taking into account the previous lemma , it follows that (from (5.2)) :

$$\deg c_1 \nu(\delta, E^\bullet) + \deg [E^\bullet]_{|\delta} = 1 \ .$$

Moreover, the inclusions $\delta \subset I \subset E^\bullet$ yield $\nu(\delta, E^\bullet) = \nu(\delta, I) + \nu(I, E^\bullet)_{|\delta}$. But $\nu(\delta, I) \simeq T\delta \oplus T\delta$ (recall that δ is the small diagonal of $I \simeq \mathbb{P}^1 \times \mathbb{P}^1 \times \mathbb{P}^1$). Also, as said already, $\nu(I, E^\bullet)$ is trivial since I is the fiber at 0 of $E^\bullet \longrightarrow V$. Since $\delta \simeq \mathbb{P}^1$, it follows that : $\deg [E^\bullet]_{|\delta} = 1 - \deg c_1 \nu(\delta, E^\bullet) = 1 - 4 = -3$. But one has (see (5.5)) $[E^\bullet]_{|I} = \alpha(A + B + C)$. Also, one sees that $[\delta] = AB + BC + CA$ in $\mathrm{CH}^2(I)$. Indeed, one can see that δ is the intersection of the two diagonals Δ_{12} and Δ_{13} in $\mathbb{P}^1 \times \mathbb{P}^1 \times \mathbb{P}^1$, i.e. $[\delta] = (A + B)(B + C) = AB + BC + CA$, from (5.4). It follows that, again from (5.4) :

$$-3 = \deg([E^\bullet]_{|\delta}) = \deg(([E^\bullet]_{|I})_{|\delta}) = \deg(\alpha(A + B + C)(AB + BC + CA)) = 3\alpha \quad .$$

Thus, $\alpha = -1$ and one has therefore proved the following lemma :

Lemma 7 In $\mathrm{CH}^1(I)$, the following equality holds :

$$[E^\bullet]_{|I} = -(A + B + C) \quad .$$

Equality (5.3) yields immediately the total Chern class in $\mathrm{CH}^1(I)$:

$$c\nu(v) = \mathbf{1} - (A + B + C) \quad . \tag{5.9}$$

5.2 Computation of $c\nu(w)_{|I}$

a) Some preliminary computations are needed. We use again the notation of § 3.1 : let X be a smooth subvariety of a smooth variety Z and let

$$P_3 : \widehat{H^3(Z)} \to Z \qquad \text{and} \qquad P_{12} : \widehat{H^3(Z)} \to \widehat{H^2(Z)}$$
$$\widehat{T} \mapsto p_3 \qquad\qquad\qquad \widehat{T} \mapsto \widehat{D}$$

be the two morphisms, where $\widehat{D} = (d_{12}, p_1, p_2)$.

Proposition 2 *Using the above notation, one has an exact sequence of vector bundles on $\widehat{H^3(X)}$:*

$$0 \longrightarrow P_3^*\nu(X,Z) \otimes \mathcal{O}(-\overline{E}) \longrightarrow \nu(\widehat{H^3(X)}, \widehat{H^3(Z)}) \longrightarrow P_{12}^*\nu(\widehat{H^2(X)}, \widehat{H^2(Z)}) \longrightarrow 0$$

where $\overline{E} = E_{23} + E_{31} + E^\bullet$ and ν denotes the normal bundle.

In order to show this proposition, we need an intermediate lemma. First, introduce the notation :

Notation 20 : If Ω is a smooth variety, its tangent bundle is denoted by $T\Omega$.

As $\widehat{H^3(X)} \subset P_{12}^{-1}(\widehat{H^2(X)})$, the differential $dP_{12} : T\widehat{H^3(Z)} \longrightarrow P_{12}^*T\widehat{H^2(Z)}$ gives by restriction a morphism

$$dP_{12|T\widehat{H^3(X)}} : T\widehat{H^3(X)} \longrightarrow P_{12}^*T\widehat{H^2(X)} .$$

A "normal derivative" morphism of vector bundles

$$d^\nu P_{12} : \nu(\widehat{H^3(X)}, \widehat{H^3(Z)}) \longrightarrow P_{12}^*\nu(\widehat{H^2(X)}, \widehat{H^2(Z)})$$

follows.

Similarly, the differential $dP_3 : T\widehat{H^3(Z)} \longrightarrow P_3^*TZ$ gives, as above, a "normal derivative" :

$$d^\nu P_3 : \nu(\widehat{H^3(X)}, \widehat{H^3(Z)}) \longrightarrow P_3^*\nu(X,Z) \quad ,$$

since $\widehat{H^3(X)} \subset P_3^{-1}(X)$.

b) Then, one has the following lemma :

Lemma 8 *With the above notation,*

(i) *$d^\nu P_{12}$ is a <u>surjective</u> morphism of vector bundles on $\widehat{H^3(X)}$;*

(ii) *if K is its kernel, one has an isomorphism of vector bundles :*

$$d^\nu P_{3|K} : K \xrightarrow{\sim} P_3^*\nu(X,Z) \quad .$$

Proof:

We first give local expressions of $d^\nu P_{12}$. One can see easily that it is sufficient to consider the case where X is a surface and Z a volume. We only study the neighborhood of the most degenerate case, i.e. $\widehat{T_0} \in \widehat{H^3(X)}$ is given by

$$\widehat{T_0} = (T_0, D_{12}, D_{23}, D_{31}, p_1, p_2, p_3) = (0^2, D_0, D_0, D_0, 0, 0, 0) \quad ,$$

where 0 is a point of X, D_0 is a doublet of support 0 and 0^2 denotes the triplet of ideal $\mathcal{M}^2$, if $\mathcal{M}$ is the ideal of 0 in $\mathcal{O}_X$ (i.e. T_0 is amorphous, cf. definition 2).

Let (x, y, z) be the coordinates of the variety Z centered at 0, the equation of X being $z = 0$, so that one takes the triplet 0^2 with ideal (x^2, xy, y^2, z) and the doublet D_0 with ideal (x^2, y, z). A chart of $\widehat{H^3(Z)}$ at $\widehat{T_0}$ is given in [LB1], p. 937 :

$$(s, \, t, \, r, \, c, \, c', \, c'', \, v, \, \rho, \, \sigma) \; . \tag{5.10}$$

In this chart, the equations of $\widehat{H^3(X)}$ are $r = \rho = \sigma = 0$ so that a trivialisation of $\nu(\widehat{H^3(X)}, \widehat{H^3(Z)})$ in the neighborhood of $\widehat{T_0}$ is given by :

$$(s, \, t, \, c, \, c', \, c'', \, v)(dr, \, d\rho, \, d\sigma) \quad , \tag{5.11}$$

where dr (resp. $d\rho$, $d\sigma$) represents a tangent vector in the direction r (resp. ρ, σ).

Similarly, a chart of $\widehat{H^2(Z)}$ at $P_{12}(\widehat{T_0}) = \widehat{D_0}$ is :

$$(s, \, t, \, r, \, a, \, c, \, e) \quad . \tag{5.12}$$

It corresponds to a point p_1 of coordinates $(s, \, t, \, r)$ and a doublet D of ideal $(x^2 + ax + b, \, -y + cx + d, \, -z + ex + f)$ close to $\widehat{D_0} = (D_0, p_1)$. In this chart, $\widehat{H^2(X)}$ is given by $r = e = 0$, since $z = 0$ is the equation of X in Z. Therefore, a trivialisation of $\nu(\widehat{H^2(X)}, \widehat{H^2(Z)})$ in the neighborhood of $\widehat{D_0}$ is :

$$(s, \, t, \, a, \, c)(dr, \, de) \; . \tag{5.13}$$

Now, in the charts (5.10) and (5.12), the morphism P_{12} is given by (see the two lines before (4.23)) :

$$(s, \, t, \, r, \, c, \, c', \, c'', \, v, \, \rho, \, \sigma) \mapsto (s, \, t, \, r, \, -2s + c''v, \, c, \, -\rho - \sigma c) \; .$$

Moreover (see [LB1], p. 937), one has the equality : $e = -\rho - \sigma c$. The local expression of $d^\nu P_{12}$

$$d^\nu P_{12} : \qquad \nu(\widehat{H^3(X)}, \widehat{H^3(Z)}) \qquad \rightarrow \quad P_{12}^* \nu(\widehat{H^2(X)}, \widehat{H^2(Z)})$$
$$(s, \, t, \, c, \, c', \, c'', \, v)(dr, \, d\rho, \, d\sigma) \; \mapsto \; (s, \, t, \, c, \, c', \, c'', \, v)(dr, \, -d\rho - c\, d\sigma) \tag{5.14}$$

follows.

We have therefore proved that $d^\nu P_{12}$ is a <u>surjection</u> of vector bundles, i.e. (i) of lemma 8.

Let us now express P_3 and $d^\nu P_3$. In the same chart (5.10) of $\widehat{H^3(Z)}$ at $\widehat{T_0}$, we have seen (cf. the lines above (4.25)) that P_3 can be expressed by :

$$\begin{aligned}
P_3(s, \, t, \, r, \, c, \, c', \, c'', \, v, \, \rho, \, \sigma) &= (s + s' + s'', \, t + t' + t'', \, r + r' + r'') \\
&= (s - v(c' + c''), \, t - v(cc'' + cc' + c'^2 + c'c''), \\
&\quad r + v(c' + c'')(\rho + \sigma c + \sigma c')) \quad ,
\end{aligned}$$

from (4.27). The local expression of $d^\nu P_3$

$$d^\nu P_3 : \nu(\widehat{H^3(X)}, \widehat{H^3(Z)}) \quad \rightarrow \quad P_3^* \nu(X, Z)$$
$$(s, t, c, c', c'', v)(dr, d\rho, d\sigma) \quad \mapsto \quad (s, t, c, c', c'', v)(dr + v(c' + c'')(d\rho + (c + c')d\sigma))$$
$$(5.15)$$

follows. Let us then restrict $d^\nu P_3$ to $K = Ker(d^\nu P_{12})$. Expression (5.14) yields :

$$d^\nu P_{3|K} : (s, t, c, c', c'', v)(0, -cd\sigma, d\sigma) \mapsto (s, t, c, c', c'', v)(v(c' + c'')c'd\sigma) .$$

But $v(c' + c'')c' = 0$ is the local equation of $\overline{E} = E^\bullet + E_{31} + E_{23}$ (see (4.28)). Therefore one sees that $d^\nu P_{3|K}$ is an isomorphism of vector bundles on $\widehat{H^3(X)}$:

$$K \xrightarrow{\sim} P_3^* \nu(X, Z) \otimes \mathcal{O}(-\overline{E}) \quad ,$$

and this proves (ii) of lemma 8. (Of course, one should also construct the local expressions of $d^\nu P_{12}$ and $d^\nu P_3$ in the neighborhood of other points of $\widehat{H^3(X)}$ – we leave it to the reader). $\qquad\qquad\square$

The last step consists in applying the trivial lemma :

Lemma 9 *Let Y be a variety, D a Cartier's divisor and E, E_1, E_2 three vector bundles on Y. Let $f_1 : E \longrightarrow E_1$ and $f_2 : E \longrightarrow E_2$ be two morphisms of bundles. Assume that :*

(i) $f_2 : E \longrightarrow E_2$ is a <u>surjective</u> morphism of bundles,

(ii) $f_{1|Ker f_2}$ is an isomorphism from $Ker f_2$ onto IE_1, where $I = \mathcal{O}_Y(-D)$.

Then one has the exact sequence of bundles on Y :

$$0 \longrightarrow IE_1 \xrightarrow{u} E \xrightarrow{f_2} E_2 \longrightarrow 0 \quad ,$$

where $u = (f_{1|Ker f_2})^{-1}$.

We apply this lemma with $f_1 = d^\nu P_3$ and $f_2 = d^\nu P_{12}$. One gets the desired sequence and proposition 2 is shown.

Remark 6 : This kind of computations recovers in a simpler way the results of [LB2] or [LB3].

c) We can now compute explicitly $c\nu(w)_{|I} = c\nu(\widehat{H^3(\Gamma)}, H^3(\widehat{V \times W}))_{|I}$ when $\Gamma \subset V \times W$ is the graph of $f : V \longrightarrow W$. Of course, we will apply the previous results with $X = \Gamma$ and $Z = V \times W$.

Let us now consider the sequence of morphisms (see the diagram at the beginning of chapter 5) :

$$I \overset{\beta'}{\hookrightarrow} \widehat{H^3(\Gamma)} \overset{P_i}{\to} \Gamma \overset{j}{\hookrightarrow} V \times W \overset{pr_2}{\to} W \tag{5.16}$$

where β' and j are canonical imbeddings and $1 \leq i \leq 3$. The composition of the four morphisms is then <u>constant</u>. Moreover, the normal bundle $N = \nu(\Gamma, V \times W)$ can be identified with the bundle (of rank 3) $j^* pr_2^* TW$. Then

$$\beta'^* P_i^* N = \beta'^* P_i^* j^* pr_2^* TW \text{ is } \underline{\text{trivial}} \text{ of rank 3,} \tag{5.17}$$

which one writes as $\beta'^* P_i^* N = \mathcal{O}^3$.

From proposition 2, one has the exact sequence of vector bundles on $\widehat{H^3(\Gamma)}$:

$$0 \longrightarrow P_3^* N \otimes \mathcal{O}(-\overline{E}) \longrightarrow \nu(w) \longrightarrow P_{12}^* \nu(\widehat{H^2(\Gamma)}, H^2(\widehat{V \times W})) \longrightarrow 0 . \tag{5.18}$$

One has also the similar exact sequence on $\widehat{H^2(\Gamma)}$ (one can refer for example to [LB2] or [LB3]) :

$$0 \longrightarrow \pi_1^* N \otimes \mathcal{O}(-F) \longrightarrow \nu(\widehat{H^2(\Gamma)}, H^2(\widehat{V \times W})) \longrightarrow \pi_2^* N \longrightarrow 0 \tag{5.19}$$

where $\pi_1, \pi_2 : \widehat{H^2(\Gamma)} \longrightarrow \Gamma$ are the natural morphisms. Of course, one has $\pi_1 \circ P_{12} = P_1$ and $\pi_2 \circ P_{12} = P_2$; moreover, $P_{12}^{-1}(F) = E_{12} + E^\bullet$ (see (4.24)). Thus, by lifting the previous exact sequence (5.19), one obtains the exact sequence of vector bundles on $\widehat{H^3(\Gamma)}$:

$$0 \longrightarrow P_1^* N \otimes \mathcal{O}(-E_{12} - E^\bullet) \longrightarrow P_{12}^* \nu(\widehat{H^2(\Gamma)}, H^2(\widehat{V \times W})) \longrightarrow P_2^* N \longrightarrow 0 \tag{5.20}$$

A look at (5.18) and (5.20) enables us to give the total Chern class :

$$c\nu(w) = c(P_3^* N \otimes \mathcal{O}(-\overline{E})) \cdot c(P_1^* N \otimes \mathcal{O}(-E_{12} - E^\bullet)) \cdot c(P_2^* N) .$$

By applying β'^* to both members of this equality, one obtains from (5.17) :

$$c\nu(w)_{|I} = c(\mathcal{O}^3 \otimes \beta'^* \mathcal{O}(-\overline{E})) \cdot c(\mathcal{O}^3 \otimes \beta'^* \mathcal{O}(-E_{12} - E^\bullet)) \cdot \mathbf{1} ,$$

which can also be expressed by

$$c\nu(w)_{|I} = \beta'^*((1 - E_{23} - E_{31} - E^\bullet)^3 (1 - E_{12} - E^\bullet)^3) , \tag{5.21}$$

since $\overline{E} = E_{23} + E_{31} + E^\bullet$.

5.3   Computation of the contribution of I

From what we have seen at the beginning of chapter 5, the contribution of I in the cycle $[\widehat{H^3(\Gamma)}] \cdot [\widehat{H^3(V)}] \times W$ is $c_3(\nu(w)_{|I} - \nu(v))$. From (5.21) and (5.3), one knows the total Chern class

$$c(\nu(w)_{|I} - \nu(v)) = \frac{\beta'^*((1 - E_{23} - E_{31} - E^\bullet)^3 (1 - E_{12} - E^\bullet)^3)}{1 + E^\bullet} , \tag{5.22}$$

where $\beta' : I \hookrightarrow \widehat{H^3(\Gamma)}$ is the canonical imbedding. Moreover (lemma 7), one has $\beta'^* E^\bullet = -(A + B + C)$ where A, B, $C \in CH^1(I) \simeq \mathbb{Z}^3$ were defined in notation 19.

Furthermore, the divisors E_{12}, E_{23}, E_{31} of $\widehat{H^3(\Gamma)}$ give, when restricted to I, the three diagonals of $\mathbb{P}^1 \times \mathbb{P}^1 \times \mathbb{P}^1$ respectively. E_{12} is for example defined as the set of $\hat{T} = (T, D_{12}, D_{23}, D_{31}, p_1, p_2, p_3)$ such that $D_{23} = D_{31}$. If one denotes by Δ_{ab}, Δ_{bc}, Δ_{ca} the three diagonals of $\mathbb{P}^1 \times \mathbb{P}^1 \times \mathbb{P}^1$ (defined by $\Delta_{ab} = \{\,(a, b, c) \mid a = b\}$, etc ...), one sees that $\beta'^* E_{12} = \Delta_{bc}$, since $D_{23} = D_{31}$.

Moreover, $\Delta_{bc} = B + C$ in $CH^1(\mathbb{P}^1 \times \mathbb{P}^1 \times \mathbb{P}^1)$ and similarly, by circular permutation :

$$\beta'^* E_{23} = \Delta_{ac} = A + C \ \in CH^1(\mathbb{P}^1 \times \mathbb{P}^1 \times \mathbb{P}^1)$$
$$\beta'^* E_{31} = \Delta_{ab} = A + B \ \in CH^1(\mathbb{P}^1 \times \mathbb{P}^1 \times \mathbb{P}^1) \ .$$

Then one has the total Chern class in $CH^\bullet(I)$:

$$c(\nu(w)_{|I} - \nu(v)) = \frac{(1 - A)^3 (1 + A)^3}{1 - (A + B + C)} \quad .$$

Since $A^2 = B^2 = C^2 = 0$ and $\deg(ABC) = 1$, it follows immediately that :

$$\deg c_3(\nu(w)_{|I} - \nu(v)) = 6 \tag{5.23}$$

Therefore, the following theorem has been proved :

Theorem 5 *Let* $f : V \longrightarrow W$ *be a morphism of a smooth surface in a smooth volume. The morphism* f *is assumed to have a* S_2*-singularity at the point* $0 \in V$*. Let* $\Gamma \subset V \times W$ *be the graph of* f*.*

Let I *be the excess component, of dimension 3, of the intersection* $\widehat{H^3(\Gamma)} \cap (\widehat{H^3(V)} \times W)$ *consisting of the complete triples* $\hat{T} = (0^2, D_{12}, D_{23}, D_{31}, 0, 0, 0)$ *where* $supp(D_{ij}) = \{0\}$*.*

The component I *is assumed to be reduced. Then, the contribution of* I *in the 0-cycle*

$$M_3 = [\widehat{H^3(\Gamma)}] \cdot [\widehat{H^3(V)} \times W]$$

is of degree 6. Therefore, the contribution of I *is 6 in the 0-cycle* $\widehat{m^3} \in CH^2(V)$*.*

Part two

Construction of a complete quadruples variety

The present goal is to construct a "good" space of ordered quadruples of a variety V, in order to give a definition and a computation of the quadruple class m_4 for an arbitrary morphism $f : X \longrightarrow Y$ between non-singular varieties (with dim $X <$ dim Y). We saw in the introduction (in 0.7) that a "naïve" generalization $\widehat{H^4_{na\"ive}}(V)$ of the construction of $\widehat{H^3(V)}$ is not sufficient : the variety $\widehat{H^4_{na\"ive}}(V)$ obtained in this way is in fact reducible and singular. Therefore, we had to construct an intermediate variety $B(V)$, which is the closure of the graph of the residual rational map (see definitions 3 and 4) :

$$Res : \quad I(V) \quad \cdots \rightarrow \quad Hilb^2(V)$$
$$d \subset q \quad \cdots \rightarrow \quad d' = q \setminus d$$

The following chapter is devoted to the study of this auxiliary variety $B(V)$. The construction of our complete quadruples variety will be given in chapter 7.

Chapter 6

Construction of the variety $B(V)$

6.1 Statement of the theorem

Theorem 6 *Let V be a non-singular, irreducible variety of dimension $\dim V \geq 3$ over $\mathbb{C}$.*

Let $B(V)$ be the closure of the graph of the residual rational map :

$$
\begin{aligned}
Res: \quad I(V) \quad &\cdots \rightarrow \quad Hilb^2(V) \\
(d, q) \quad &\cdots \rightarrow \quad d' = q \setminus d
\end{aligned}
$$

where $I(V) \subset Hilb^2(V) \times Hilb^4(V)$ is the incidence variety.

The variety $B(V)$ is irreducible and non-singular of dimension $4 \cdot \dim V$.

The irreducibility of the variety $B(V)$ will be established in § 6.3. The proof of the non-singularity of the variety $B(V)$ will be the subject of § 6.4. One can go back to the case where V is a variety of dimension 3 in a systematic manner. When $n = \dim V \geq 4$, one just has to replace z by $z_1, ..., z_{n-2}$ everywhere in the computations.

In the following, V will be a non-singular variety of dimension 3.

Let us give some definitions :

6.2 Definitions, drawing conventions

Definitions 7 : Let q be a quadruplet of support a closed point p of V. According to the description of Briançon [B1], the different quadruplets supported by p are given by the different ideals of $\mathcal{O}_V$ (in an appropriate coordinate system (x, y, z) centered at p) :

(i) $I(q) = (x^4, y, z)$

This quadruplet is called a curvilinear quadruplet.

(ii) $I(q) = (x^2, y^2, z)$

This quadruple point is called a square quadruplet .

(iii) $I(q) = (x^3, xy, y^2, z)$

This quadruplet is said to be elongated.

(iv) $I(q) = (x, y, z)^2$

This quadruple point is said to be spherical.

Drawing conventions :

One will use the following drawing conventions :

– The following symbol will represent a curvilinear quadruple point :

– The following symbol will be used to represent a square quadruplet :

– An elongated quadruple point will be represented by the following drawing :

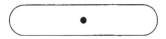

– Finally, the following symbol will represent a spherical quadruplet :

6.3   Irreducibility and dimension of $B(V)$

6.3.1   General facts on Hilbert schemes :

Here, we recall some generalities on Hilbert schemes :

Property 1 : Universal property of the Hilbert scheme

The Hilbert scheme $Hilb^d(V)$ comes equipped with a d-sheeted tautological cover-ing, denoted by $'Hilb^d(V)$, and defined in the product $V \times Hilb^d(V)$. ϖ denotes the projection from $'Hilb^d(V) \subset V \times Hilb^d(V)$ on $Hilb^d(V)$; the projection ϖ (called *universal family*) is flat by definition. From a set-theoretic point of view, $'Hilb^d(V)$ contains the couples (p, ξ) such that the point p is a subscheme of ξ.

The Hilbert scheme $Hilb^d(V)$ is solution of the following universal problem :

Let S be a scheme. Let $\mathcal{Y} \subset S \times V$ be a flat d-sheeted ramified cover of S (via the first projection). Giving such a subscheme $\mathcal{Y}$ of $S \times V$ is equivalent to giving a unique morphism $f : S \to Hilb^d(V)$. The family $\mathcal{Y}$ is obtained by the *pull-back* of the universal family $'Hilb^d(V)$ by f.

One recalls the <u>flatness criterion</u> for a <u>finite</u> morphism :

Let $\varphi : X \to T$ be a finite morphism of schemes, with T integral. Then φ is flat over T if and only if the length of the fibers $\varphi^{-1}(t)$ is a constant d independent of $t \in T$.

Property 2 : Recall that if $\xi \in Hilb^d(V)$ is a d-uple union of ξ_1 and ξ_2 where ξ_1 is a d_1-uple of support p_1 and ξ_2 is a $(d-d_1)$-uple disjoint from p_1, then $Hilb^d(V)$ is locally isomorphic at ξ to the product $Hilb^{d_1}(V_1) \times Hilb^{d-d_1}(V_2)$, where V_1 is a neighborhood of V at ξ_1 and V_2 is a neighborhood of V at ξ_2.

Improperly, we will say that the d-uple ξ_ε is deforming to ξ in $Hilb^d(V)$ and we will denote this deformation by $\xi_\varepsilon \to \xi$ when ε goes to 0, if the family $(\xi_\varepsilon)_{\varepsilon \in \mathbb{C}}$ defined in this way corresponds to a sub-family of $V \times \mathbb{C}$, flat over $\mathbb{C}$, via the second projection. Said differently, this deformation of base $\mathbb{C}$ corresponds to a unique morphism from $\mathbb{C}$ to $Hilb^d(V)$.

6.3.2

Recall that $I(V)$ is the incidence subvariety of $Hilb^2(V) \times Hilb^4(V)$ consisting of the elements (d, q) such that d is a subscheme of q. Also recall that Π_2 denotes the projection from $I(V)$ onto $Hilb^4(V)$. One introduces some new notation :

Notation 21 : From now on, one denotes by $H^d(V)$ the Hilbert scheme $Hilb^d(V)$ of the d-uples of V.

Notation 22 :

- For $d \leq 4$, we denote by $H_{\neq}^d(V)$ the dense open subset [F] of $H^d(V)$ containing the d-uples formed by d simple points.

- One denotes by $I_{\neq}(V)$ the open subset of $I(V)$ containing the elements (d, q) such that $q \in H^4_{\neq}(V)$.

- One denotes by $\widetilde{H^3(V)}$ the subvariety of $H^2(V) \times H^3(V)$ containing the (d, t) such that d is a subscheme of t. This subvariety is non-singular [ELB]. The second projection

$$\widetilde{H^3(V)} \rightarrow H^3(V)$$
$$(d, t) \mapsto t$$

is generically a 3-sheeted covering.

The Hilbert scheme $H^4(V)$ has a natural stratification consisting of five strata denoted by $H^4_4(V)$, $H^4_{22}(V)$, $H^4_{31}(V)$, $H^4_{211}(V)$ and $H^4_{1111}(V)$.

- The stratum $H^4_4(V)$ is the closed subvariety of $H^4(V)$ containing quadruple points (i.e. the support is only one point).

- The stratum $H^4_{22}(V)$ is the locally closed subvariety of $H^4(V)$ containing quadruplets which are the union of two double points.

- The stratum $H^4_{31}(V)$ is the locally closed subvariety of $H^4(V)$ containing quadruplets which are the union of a triple point and a simple point.

- The stratum $H^4_{211}(V)$ is the locally closed subvariety of $H^4(V)$ containing quadruplets which are the union of a double point and two simple points.

- The stratum $H^4_{1111}(V)$ is the open subset containing simple quadruplets, previously denoted by $H^4_{\neq}(V)$.

This natural stratification of $H^4(V)$ will induce through the projection $\Pi_2 : I(V) \rightarrow H^4(V)$ a stratification on $I(V)$, denoted by $I_{\bullet}(V)$. Note that the open subset $I_{1111}(V)$ of $I(V)$ has already been denoted by $I_{\neq}(V)$.

Other drawing conventions :

– We recall the drawing convention used to represent the double point d of support a point p of V :

– We will represent a n-uple curvilinear point (i.e. a subscheme of a non-singular curve) by the symbol :

– Then, for an amorphous triplet t (cf. definition 2), we will use the following convention :

Our goal now is to prove the following proposition :

Proposition 3 *The incidence subvariety $I(V)$ of $H^2(V) \times H^4(V)$ is irreducible of dimension $12 = 4 \cdot dim(V)$.*

Proof :

We will show that the open subset $I_{\neq}(V) \subset I(V)$ is dense in $I(V)$. According to property 2, it is enough to prove that each element (d, q) of the stratum $I_4(V)$ (i.e. when the support of the quadruplet q is one point) is the limit of elements of $I_{\neq}(V)$. In fact, when the support of the quadruplet contains at least two points, the result is already known, as shown below :

a) In an element of the stratum $I_{31}(V)$:

With our drawing conventions, the elements of this stratum are of one of the four following forms :

– The quadruplet q is the union of a triple point t and a simple point m, the doublet d is simple :

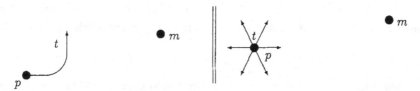

Figure 6.1: $d = p \cup m$ and $q = t \cup m$

– The quadruplet q is the union of a triple point t and a simple point m, the doublet d is contained in the triplet t :

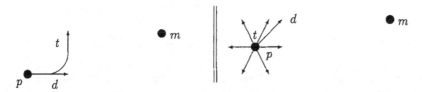

Figure 6.2: $d \subset t$ and $q = t \cup m$

From property 2, the Hilbert scheme $H^4(V)$ is locally isomorphic at q to the product $H^3(V) \times V$. If the doublet d is simple (figure 6.1), the variety $I(V)$ is locally isomorphic to the product $'H^3(V) \times V$ which is an irreducible variety of dimension 12. If the doublet d is a double point (figure 6.2), the incidence variety is in this case locally isomorphic to the product $\widetilde{H^3(V)} \times V$ which is a variety of dimension 12. Moreover, each element of this stratum can be obtained as the limit of elements of the open subset $I_{\neq}(V)$.

b) In an element of the stratum $I_{22}(V)$:
The quadruplet q is the union of two double points d_1 and d_2 of support p_1 and p_2 :

Figure 6.3: $q = d_1 \cup d_2$

Still from property 2, the Hilbert scheme $H^4(V)$ is locally isomorphic at q to the product $H^2(V) \times H^2(V)$. If d is the union of the two simple points p_1 and p_2, the incidence variety is locally isomorphic to the product $'H^2(V) \times 'H^2(V)$. If d is one of the two doublets d_1, d_2, the variety $I(V)$ is then locally isomorphic to the product $H^2(V) \times H^2(V)$. In these two cases, $I(V)$ is a variety of dimension 12 and the elements of this stratum can be obtained again as the limits of elements of the open set $I_{\neq}(V)$.

c) In an element of the stratum $I_{211}(V)$:
The quadruplet q is the union of a double point d_1 of support p_1 and two simple points p_2 and p_3 :

Figure 6.4: $q = d_1 \cup p_2 \cup p_3$

If $d = d_1$ or $d = p_2 \cup p_3$, the incidence variety is in this case locally isomorphic to the product $H^2(V) \times H^2(V)$ which is a variety of dimension 12. Now, if d is one of the two simple doublets $p_1 \cup p_2$, $p_1 \cup p_3$, the variety $I(V)$ is then locally isomorphic to the product $V \times V \times H^2(V)$ which is of dimension 12. The points of this stratum belong to the closure of $I_{\neq}(V)$ in $I(V)$.

d) It remains to study the elements of the stratum $I_4(V)$:

Let us denote by p the support of an element (d, q) of $I_4(V)$. Remember that Π_2 is the projection from $I(V)$ onto $H^4(V)$.

(i) If q is a <u>curvilinear</u> quadruple point, there is only one element (d, q) in the fiber $\Pi_2^{-1}(q)$, where d is the only doublet contained in q. In an appropriate local coordinate system (x, y, z) centered at p (cf. definition 7.(i)), the quadruplet q is defined by the ideal $I(q) = (x^4, y, z)$ of $\mathcal{O}_V$ and the doublet d has for ideal $I(d) = (x^2, y, z)$.

Let us consider the ideal $I_\varepsilon = (x^4 - \varepsilon^4, y, z)$; it is the ideal of a quadruplet q_ε of $H^4_{\neq}(V)$.

The ideal $J_\varepsilon = (x^2 - \varepsilon^2, y, z)$ defines a doublet d_ε of $H^2_{\neq}(V)$. Since the inclusion of ideals $I_\varepsilon \subset J_\varepsilon$ is equivalent to the scheme-theoretic inclusion $d_\varepsilon \subset q_\varepsilon$, it follows that for every ε different from zero, the element $(d_\varepsilon, q_\varepsilon)$ is in $I_{\neq}(V)$. In addition, $I_\varepsilon = I(q_\varepsilon)$ obviously goes to $I(q)$ when ε goes to 0. Similarly, $J_\varepsilon = I(d_\varepsilon)$ goes to $I(d)$ when ε goes to 0.

So, when the quadruplet q is curvilinear, the element (d, q) of $I(V)$ is the limit of elements of $I_{\neq}(V)$.

(ii) If q is <u>square</u>, from definition 7.(ii), one can assume it to be defined by the ideal $I(q) = (x^2, y^2, z)$ of $\mathcal{O}_V$, where (x, y, z) is an appropriate local coordinate system centered at p. As the coordinates x and y play a symmetric role, one can always assume that a doublet d^α which constitutes an element (d^α, q) of the fiber $\Pi_2^{-1}(q)$ is given by the ideal $I(d^\alpha) = (x^2, y - \alpha x, z)$, where α is a fixed scalar.

For $\alpha \neq 0$, the ideal $I_\varepsilon = (x(x - \varepsilon), y(y - \alpha\varepsilon), z)$ defines the quadruplet q_ε which is the union of the four following simple points :

$$
p_{1\varepsilon}\begin{vmatrix} 0 \\ 0 \\ 0 \end{vmatrix} \quad p_{2\varepsilon}\begin{vmatrix} \varepsilon \\ 0 \\ 0 \end{vmatrix} \quad p_{3\varepsilon}\begin{vmatrix} \varepsilon \\ \alpha\varepsilon \\ 0 \end{vmatrix} \quad p_{4\varepsilon}\begin{vmatrix} 0 \\ \alpha\varepsilon \\ 0 \end{vmatrix}
$$

(The notation $m\begin{vmatrix} a \\ b \\ c \end{vmatrix}$ represents the coordinates in $\mathbb{C}^3$ of the point m. The point m is then defined by the ideal $(x - a, y - b, z - c)$.)

Obviously, $I(q_\varepsilon)$ goes to $I(q)$ when ε goes to 0. The ideal $(x(x - \varepsilon), y - \alpha x, z)$ defines the doublet d_ε which is the union of the two simple points $p_{1\varepsilon}$ and $p_{3\varepsilon}$. The doublet d_ε deforms in d^α in $H^2(V)$ because $I(d_\varepsilon)$ goes to $I(d^\alpha)$ when ε goes to 0. We represent these different configurations as follows :

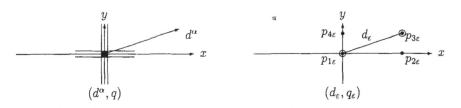

$$(d^\alpha, q) \qquad\qquad\qquad (d_\varepsilon, q_\varepsilon)$$

If $\alpha = 0$, the element (d°, q) of $I_4(V)$ is the limit of elements $(d_\varepsilon, q_\varepsilon)$ of $I_{\neq}(V)$, where q_ε is the quadruplet union of the following four simple points :

$$p_{1\varepsilon} \begin{array}{|c}0\\0\\0\end{array} \quad p_{2\varepsilon} \begin{array}{|c}\varepsilon\\0\\0\end{array} \quad p_{3\varepsilon} \begin{array}{|c}\varepsilon\\\varepsilon\\0\end{array} \quad p_{4\varepsilon} \begin{array}{|c}0\\\varepsilon\\0\end{array}$$

The quadruplet q_ε is defined by the ideal $I(q_\varepsilon) = (x(x - \varepsilon), y(y - \varepsilon), z)$ and it goes to $I(q)$ when ε goes to 0. If d_ε is the simple doublet $p_{1\varepsilon} \cup p_{2\varepsilon}$, d_ε is defined by the ideal $I(d_\varepsilon) = (x(x - \varepsilon), y, z)$ which goes to $I(d^\circ)$ when ε goes to 0. Again, we represent these configurations as follows :

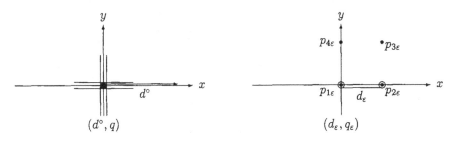

$$(d^\circ, q) \qquad\qquad\qquad (d_\varepsilon, q_\varepsilon)$$

So, if q is a square quadruplet, each element (d, q) in $I_4(V)$ is the limit of elements of $I_{\neq}(V)$.

(iii) Now, if the quadruplet q is <u>elongated</u>, from definition $7.(iii)$, one can assume q to be defined by the ideal $I(q) = (x^3, xy, y^2, z)$. A doublet d^α which constitutes an element (d^α, q) of the fiber $\Pi_2^{-1}(q)$ is the same as the one given by an ideal $I(d^\alpha) = (x^2, y - \alpha x, z)$.

For $\alpha \neq 0$, one considers the quadruplet q_ε which is the union of the four following simple points :

$$p_{1\varepsilon} \begin{array}{|c}0\\0\\0\end{array} \quad p_{2\varepsilon} \begin{array}{|c}\varepsilon\\0\\0\end{array} \quad p_{3\varepsilon} \begin{array}{|c}\varepsilon\\\alpha\varepsilon\\0\end{array} \quad p_{4\varepsilon} \begin{array}{|c}-\varepsilon\\0\\0\end{array}$$

This quadruplet is defined by the ideal :

$$\begin{aligned}I(q_\varepsilon) &= (x(x^2 - \varepsilon^2), y, z) \cap (x - \varepsilon, y - \alpha\varepsilon, z)\\&= (x(x^2 - \varepsilon^2), y(x - \varepsilon), y(y - \alpha\varepsilon), z)\end{aligned}$$

This quadruplet q_ε goes to q in $H^4(V)$ when ε goes to 0. If d_ε is the doublet union of the two simple points $p_{1\varepsilon}$ and $p_{3\varepsilon}$, d_ε is defined by $I(d_\varepsilon) = (x(x-\varepsilon), y - \alpha x, z)$. This ideal clearly goes to $I(d^\alpha)$ when ε goes to 0. Let us draw below these configurations :

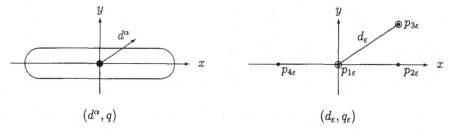

$$(d^\alpha, q) \qquad\qquad\qquad (d_\varepsilon, q_\varepsilon)$$

If $\alpha = 0$, the element (d°, q) of $I_4(V)$ is the limit of elements $(d_\varepsilon, q_\varepsilon)$ of $I_{\neq}(V)$, where q_ε is the quadruplet which is the union of the four following simple points :

$$
p_{1\varepsilon}\begin{vmatrix} 0 \\ 0 \\ 0 \end{vmatrix} \quad
p_{2\varepsilon}\begin{vmatrix} \varepsilon \\ 0 \\ 0 \end{vmatrix} \quad
p_{3\varepsilon}\begin{vmatrix} 0 \\ \varepsilon \\ 0 \end{vmatrix} \quad
p_{4\varepsilon}\begin{vmatrix} -\varepsilon \\ 0 \\ 0 \end{vmatrix}
$$

This quadruplet is defined by the ideal $I(q_\varepsilon) = (x(x^2 - \varepsilon^2), yx, y(y-\varepsilon), z)$. The simple doublet $d_\varepsilon = p_{1\varepsilon} \cup p_{2\varepsilon}$ is defined by the ideal $I(d_\varepsilon) = (x(x-\varepsilon), y, z)$. Here are the configurations :

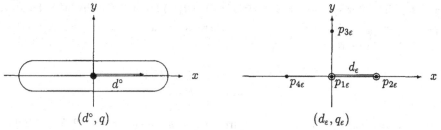

$$(d^\circ, q) \qquad\qquad\qquad (d_\varepsilon, q_\varepsilon)$$

So, when the quadruplet q is elongated, every element (d, q) of $I(V)$ belongs to $\overline{I_{\neq}(V)}$, the closure of $I_{\neq}(V)$ in $I(V)$.

(iv) If the quadruplet q is spherical, that is to say defined by the ideal m_p^2 where m_p is the maximal ideal of $\mathcal{O}_V$, the fiber $\Pi_2^{-1}(q)$ is isomorphic to $\mathbb{P}(T_p V)$ (the projective space associated to the vector tangent space of V at p). As there is no preferred direction in such a quadruplet, it is sufficient to prove that one of the elements (d, q) of the fiber $\Pi_2^{-1}(q)$ is in the closure of $I_{\neq}(V)$ in order to obtain the result for all the elements of the fiber.

For an appropriate local coordinate system (x, y, z) centered at p, the doublet d is defined by the ideal (x^2, y, z). The element (d, q) is the limit of elements $(d_\varepsilon, q_\varepsilon)$ of $I_{\neq}(V)$, where q_ε is the simple quadruplet $q_\varepsilon = p_{1\varepsilon} \cup p_{2\varepsilon} \cup p_{3\varepsilon} \cup p_{4\varepsilon}$ and d_ε is the simple doublet $d_\varepsilon = p_{1\varepsilon} \cup p_{2\varepsilon}$. The coordinates of the points are :

$$
\begin{array}{c|c} & 0 \\ p_{1\varepsilon} & 0 \\ & 0 \end{array}
\qquad
\begin{array}{c|c} & \varepsilon \\ p_{2\varepsilon} & 0 \\ & 0 \end{array}
\qquad
\begin{array}{c|c} & 0 \\ p_{3\varepsilon} & \varepsilon \\ & 0 \end{array}
\qquad
\begin{array}{c|c} & 0 \\ p_{4\varepsilon} & 0 \\ & \varepsilon \end{array}
$$

Again, one represents these configurations as :

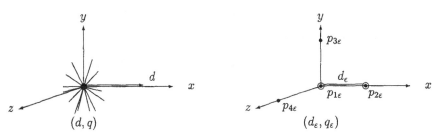

$$(d, q) \qquad\qquad\qquad\qquad (d_\varepsilon, q_\varepsilon)$$

The quadruplet q_ε is defined by the ideal $I(q_\varepsilon) = (x(x-\varepsilon), xy, xz, y(y-\varepsilon), yz, z(z-\varepsilon))$, which goes to $I(q)$ when ε goes to 0. The doublet d_ε of ideal $I(d_\varepsilon) = (x(x-\varepsilon), y, z)$ goes to d in $H^2(V)$ when ε goes to 0.

So, when q is a spherical quadruplet, the element (d, q) of $I(V)$ is in $\overline{I_{\neq}(V)}$.

We have therefore shown the inclusion of the stratum $I_4(V)$ in $\overline{I_{\neq}(V)}$. As it is the same for the other strata, the equality $\overline{I_{\neq}(V)} = I(V)$ follows. On the other hand, the open subset $I_{\neq}(V)$ is irreducible because there is a birational morphism from $I(V)$ to the irreducible product $H^2(V) \times H^2(V)$:

$$
\begin{array}{ccc}
I(V) & \cdots\to & H^2(V) \times H^2(V) \\
(d, q) & \cdots\to & (d, d' = q \setminus d)
\end{array}
$$

The incidence variety $I(V)$ is therefore irreducible of dimension $12 = 2 \cdot \dim (H^2(V))$. So, we have proved proposition 3.

On the other hand, the projection

$$
\begin{array}{cccc}
\Pi_2 : & I(V) & \to & H^4(V) \\
& (d, q) & \mapsto & q
\end{array}
$$

is generically a $6 = \begin{pmatrix} 4 \\ 2 \end{pmatrix}$-sheeted covering because the open set $H^4_{\neq}(V)$ is dense in $H^4(V)$.

Then proposition 4 follows trivially :

Proposition 4 *The closure $B(V)$ of the graph of the residual rational map Res is an irreducible subvariety of $I(V) \times H^2(V)$ of dimension 12.*

Proof :

Remember that $U \subset I(V)$ is the open subset of $I(V)$ where the rational residual map Res is regular. The graph Γ_{Res_U} of Res restricted to U is isomorphic to U, which is a dense open set of dimension 12, from proposition 3. The closure $B(V)$ of Γ_{Res_U} in $I(V) \times H^2(V)$ is then irreducible of dimension 12. □

6.4 Non-singularity of $B(V)$

Remember that Π denotes the projection :

$$\Pi : \quad R(V) \quad \to \quad H^4(V)$$
$$(d, q, d') \quad \mapsto \quad q$$

and π its restriction to $B(V)$. Also remember that $\tilde{q}$ denotes an element of the fiber $\pi^{-1}(q) \subset B(V)$, (see § 0.7, definition 4 and notation 2). The study of the non-singularity of $B(V)$ reduces to the cases where the support of the element $\tilde{q} \in B(V)$ consists of exactly one point. When the support of the quadruplet q consists of at least two points, the variety $B(V)$ is in fact locally isomorphic at $\tilde{q}$ to a smooth variety of dimension $12 = 4 \cdot \dim (V)$:

- If the quadruplet q is the union of a triple point t of support p and a simple point m, according to property 2 the variety $B(V)$ is locally isomorphic at $\tilde{q} = \widetilde{t \cup m}$ to the product $\widetilde{H^3(V)} \times V$, where $\widetilde{H^3(V)}$ denotes the incidence subvariety of $H^2(V) \times H^3(V)$ (cf. notation 22),

- When the quadruplet q is the union of two double points d_1 and d_2 of distinct supports, again from property 2, the variety $B(V)$ is locally isomorphic at $\tilde{q} = \widetilde{d_1 \cup d_2}$ to the product $'H^2(V) \times 'H^2(V)$.

In these two cases, the result is a smooth variety (of dimension 12) since it is locally the product of two smooth varieties.

The goal of § 6.4.1 is to prove the non-singularity of $B(V)$ at every element $\tilde{q}_o$, where q_o is a locally complete intersection quadruple point (i.e. q_o is curvilinear or square). Remember that the complete intersection k-uplets ξ are smooth points of the Hilbert scheme $H^k(V)$. It results from [H]-II-prop. 8.21.A(e). We will then prove in § 6.4.2 the non-singularity of $B(V)$ at the points $\tilde{q}_o$ where q_o is a non-locally complete intersection quadruple point (i.e. q_o elongated or spherical) . Remember [I2, F] that the Hilbert scheme $H^4(V)$ is irreducible and singular at the spherical quadruplets q, that is to say defined by the ideal $\mathcal{M}_p^2$, where $\mathcal{M}_p$ is the ideal of a closed point p of V. (Also remember that we have assumed $\dim (V) = 3$.)

In this whole section, the support of the quadruple point q_o is denoted by p. We denote by (x, y, z) an appropriate local coordinate system centered at p (cf. definitions 7), i.e. a system in which the quadruplet q_o is defined by the ideal :

(i) (x^4, y, z), if q_o is curvilinear,

(ii) (x^2, y^2, z), if q_o is square,

(iii) (x^3, xy, y^2, z), if q_o is elongated,

(iv) $(x, y, z)^2$, if q_o is spherical.

6.4.0   Preliminaries

a) In the following, we will divide elements of the algebra $\mathbb{C}\{x, y, z\}$ by ideals of $\mathbb{C}\{x, y, z\}$. In [B2], Briançon gives a generalization of the division theorem of Weierstrass, according to the method of Hironaka (generalization in the sense that we divide a germ of an analytic function by <u>several</u> others, with respect to several coordinates). Let us now give some details on this division theorem. For each multi-index $\alpha = (\alpha_1, \alpha_2, \alpha_3)$, the notation $(xyz)^\alpha$ denotes the monomial $x^{\alpha_1} y^{\alpha_2} z^{\alpha_3}$. Recall some definitions :

Definitions 8 :
• A non zero linear form with positive integer coefficients is called direction L. For each element f in $\mathbb{C}\{x, y, z\}$, let

$$f = \sum_{\alpha \in \mathbb{N}^3} a_\alpha \cdot (xyz)^\alpha \quad .$$

• The set

$$N(f) = \{\alpha \in \mathbb{N}^3 \mid a_\alpha \neq 0\}$$

is called Newton's diagram of f.
• The integer

$$d_L(f) = inf\{L(\alpha) \mid a_\alpha \neq 0\}$$

is called L-graduation of f.
• The element of $\mathbb{C}[x, y, z]$:

$$in_L(f) = \sum_{L(\alpha) = d_L(f)} a_\alpha \cdot (xyz)^\alpha$$

is called initial form of f with respect to the direction L.

Such a positive linear form L defines an order on $\mathbb{N}^3$, denoted by $<$:

$$\alpha = (\alpha_1, \alpha_2, \alpha_3) < \beta = (\beta_1, \beta_2, \beta_3)$$

if and only if :
- either $L(\alpha_1, \alpha_2, \alpha_3) < L(\beta_1, \beta_2, \beta_3)$,

- or $L(\alpha_1, \alpha_2, \alpha_3) = L(\beta_1, \beta_2, \beta_3)$ and there exists an index i_0 such that $\alpha_{i_0} < \beta_{i_0}$ and for each index $j > i_0$, $\alpha_j = \beta_j$.

The linear form L then allows an ordering of the monomials of $\mathbb{C}\{x, y, z\}$.

Definition 9 : The smallest element of $N(f)$ for this order is called a dominant exposant of f with respect to the direction L, and it is denoted by $exp_L(f)$.

For each ideal I and for each direction L, one can associate the set of the dominant exposants $E_L(I)$ of I and the set $F_L(I) = (\alpha_1, ..., \alpha_p)$ which is the minimal finite subset of $\mathbb{N}^3$ such that $E_L(I) = \cup_{1 \le i \le p}(\alpha_i + \mathbb{N}^3)$ (see [B1], [B2], [Gal]). Let $\Delta_L(I) = \mathbb{N}^3 - E_L(I)$; there exists a unique basis $\{f_\alpha\}$ of I, called standard basis of I with respect to L indexed by $F_L(I)$, such that

$$f_\alpha(x, y, z) = (xyz)^\alpha + \sum_{\beta \in \Delta_L(I)} a_\beta (xyz)^\beta , \ \alpha \in F_L(I) \ .$$

Each element of $\mathbb{C}\{x, y, z\}$ is equivalent modulo I to a unique element of $\mathbb{C}\{x, y, z\}$ of the form :

$$\sum_{\beta \in \Delta_L(I)} a_\beta (xyz)^\beta$$

(see [B2], [Gal], thm 1.2.5 p. 124).

In particular, if I is of finite colength n, the family $\{\overline{(xyz)^\beta}\}_{\beta \in \Delta_L(I)}$ is a basis of the quotient $\dfrac{\mathbb{C}\{x, y, z\}}{I}$ over $\mathbb{C}$ and $\#\Delta_L(I) = n$.

$b)$ Moreover, we will frequently use the division theorem of Galligo by a family with parameters ([Gal], thm 1.2.7 p. 126) in the following particular case :

Let $I = (f_1, \ldots, f_p)$ be the ideal in $\mathbb{C}\{x, y, z\}$ of a n-uple point ξ of $H^n(V)$. The quotient $\dfrac{\mathbb{C}\{x, y, z\}}{(f_1, \ldots, f_p)}$ is a $\mathbb{C}$-vector space of dimension n. One assumes $(f_1, \ldots, f_p)$ to be the standard basis of I with respect a direction L ([B2], [Gal]). The monomial basis of $\dfrac{\mathbb{C}\{x, y, z\}}{(f_1, \ldots, f_p)}$ is denoted by $\{\overline{(xyz)^\alpha}\}_{\alpha \in \Delta_L(I)}$, where "$\bar{\ }$" denotes the class of an element of $\mathbb{C}\{x, y, z\}$ modulo the ideal I. Let

$$F_i(x, y, z, \underline{a}) = f_i(x, y, z) + \sum_{\alpha \in \Delta_L(I)} a_{i,\alpha}(xyz)^\alpha$$

$$\underline{a} = (a_{i,\alpha})_{1 \le i \le p, \alpha \in \Delta_L(I)}$$

One denotes by $\mathcal{O}_S$ the $\mathbb{C}$-algebra $\mathbb{C}\{\underline{a}\}$. For each element f in $\mathcal{O}_S\{x, y, z\}$, one can perform the division of f by the family $F_1, \ldots, F_p$ and the remainder h is an element of the form :

$$h(x, y, z, \underline{a}) = \sum_{\alpha \in \Delta_L(I)} h_\alpha(\underline{a})(xyz)^\alpha$$

where $h_\alpha(\underline{a}) \in \mathcal{O}_S$.

Definition 10 ([Gal], def. 1.4.5 p. 135) : Let $\varphi : X \to S$ be a morphism of germs of analytic spaces. One denotes by $\varphi^* : \mathcal{O}_S \to \mathcal{O}_X$ the corresponding local $\mathbb{C}$-algebra morphism.

There exists ([Gal], thm. 1.4.4 p. 135) a unique $\mathbb{C}$-algebra $\mathcal{O}_P$, which is the quotient of $\mathcal{O}_S$, equipped with a canonical surjection $\chi_P^* : \mathcal{O}_S \to \mathcal{O}_P$, satisfying the conditions :

(i) $\mathcal{O}_X \otimes_{\mathcal{O}_S} \mathcal{O}_P$ is a flat $\mathcal{O}_P$-module,

(ii) For each morphism of algebras $\lambda^* : \mathcal{O}_S \to \mathcal{O}_T$ such that $\mathcal{O}_X \otimes_{\mathcal{O}_S} \mathcal{O}_T$ is a flat $\mathcal{O}_T$-module, there exists one and only one morphism $\mu^* : \mathcal{O}_P \to \mathcal{O}_T$ such that $\lambda^* = \mu^* \circ \chi_P^*$.

One calls flattener of the morphism φ, the morphism $\varphi_P : X \times_S P \to P$ obtained by the change of basis $P \hookrightarrow S$.

Property 3 : One denotes by $(\mathcal{X}, 0)$ the germ of an analytic subspace of $(\mathbb{C}^3 \times \mathbb{C}^{np}, 0)$ defined by $F_1(x, y, z, \underline{a}) = \cdots = F_p(x, y, z, \underline{a}) = 0$. The natural projection from $V((F_1, \ldots, F_p)) \subset (\mathbb{C}^3 \times \mathbb{C}^{np}, 0)$ onto $(S, 0) = (\mathbb{C}^{np}, 0)$ will be denoted by Π. The Hilbert scheme $H^n(V)$ is locally isomorphic at ξ to the flattener of Π (see [Gr], prop. 0.4 p. 20). One denotes by $(P, 0)$ the basis of this flattener and $\mathcal{O}_P$ the corresponding local algebra. ($\mathcal{O}_P$ is a quotient of $\mathcal{O}_S$.)

One recalls the flatness and division theorem of Galligo ([Gal], thm 1.2.8 p. 126) which will enable us to compute explicitly local flatteners. One has the equivalent statements :

1. The module $\dfrac{\mathcal{O}_S\{x, y, z\}}{(F_1, \ldots, F_p)}$ is $\mathcal{O}_S$-flat,

2. For each f in $\mathcal{O}_S\{x, y, z\}$, the remainder of the division of f by the ideal $(F_1, \ldots, F_p)$ is zero if and only if $f \in (F_1, \ldots, F_p)$.

Remark 7 : If $I = I(\xi)$ is a complete intersection ideal, the morphism Π introduced previously is flat. Then one has $(P, 0) = (S, 0)$ – the Hilbert scheme $H^n(V)$ is smooth at ξ, of dimension $3n$.

In the following, we will use the result :

Result 1 : According to the previous property, the Hilbert scheme $H^n(V)$ is locally isomorphic at ξ to the flattener of basis P.

[$\mathcal{R}es_1$] If (f_1, f_2, f_3) is the ideal (x^2, y, z) of $\mathbb{C}\{x, y, z\}$ of a double point of $H^2(V)$, let :

$$
\begin{aligned}
F_1(x, y, z, \underline{A}) &= x^2 + ax + b \\
F_2(x, y, z, \underline{A}) &= y - cx - d \\
F_3(x, y, z, \underline{A}) &= z - ex - f \\
\underline{A} &= (a, b, c, d, e, f) \quad .
\end{aligned}
$$

One denotes by $\mathcal{O}_S$ the local algebra $\mathbb{C}\{\underline{A}\}$. For each element f in $\mathcal{O}_P\{x, y, z\}$, the remainder h of the division of f by F_1, F_2, F_3 is of the form :

$$h(\underline{A}, x, y, z) = h_0(\underline{A}) \cdot 1 + h_1(\underline{A}) \cdot x \quad,$$

where $h_0(\underline{A})$ and $h_1(\underline{A})$ are in $\mathcal{O}_P$.

[$\mathcal{R}es_2$] If (f_1, f_2, f_3) is the ideal (x^3, y, z) of $\mathbb{C}\{x, y, z\}$ of a curvilinear triple point of V, let :

$$
\begin{aligned}
F_1(x, y, z, \underline{a}) &= x^3 + a_1 x^2 + a_2 x + a_3 \\
F_2(x, y, z, \underline{a}) &= y + a_4 x^2 + a_5 x + a_6 \\
F_3(x, y, z, \underline{a}) &= z + a_7 x^2 + a_8 x + a_9 \\
\underline{a} &= (a_1, \ldots, a_9) \quad.
\end{aligned}
$$

Let $\mathcal{O}_S$ be the local algebra $\mathbb{C}\{\underline{a}\}$. For each element f in $\mathcal{O}_S\{x, y, z\}$, the remainder h of the division of f by F_1, F_2, F_3 is of the form :

$$h = h_0(\underline{a}) + h_1(\underline{a})x + h_2(\underline{a})x^2 \quad,$$

where $h_0(\underline{a}), h_1(\underline{a})$ and $h_2(\underline{a})$ are in $\mathcal{O}_S$.

[$\mathcal{R}es_3$] If (f_1, f_2, f_3, f_4) is the ideal (x^2, xy, y^2, z) of $\mathbb{C}\{x, y, z\}$ of an amorphous triplet of V, let :

$$
\begin{aligned}
F_1(x, y, z, \underline{a}) &= x^2 + a_1 x + a_2 y + a_3 \\
F_2(x, y, z, \underline{a}) &= xy + a_4 x + a_5 y + a_6 \\
F_3(x, y, z, \underline{a}) &= y^2 + a_7 x + a_8 y + a_9 \\
F_4(x, y, z, \underline{a}) &= z + a_{10} x + a_{11} y + a_{12} \\
\underline{a} &= (a_1, \ldots, a_{12}) \quad.
\end{aligned}
$$

For each element f in $\mathcal{O}_P\{x, y, z\}$, the remainder h of the division of f by F_1, F_2, F_3, F_4 is of the form :

$$h = h_0(\underline{a}) + h_1(\underline{a})x + h_2(\underline{a})y \quad,$$

where $h_0(\underline{a}), h_1(\underline{a})$ and $h_2(\underline{a})$ are in $\mathcal{O}_P$.

[$\mathcal{R}es_4$] If (f_1, f_2, f_3) is the ideal (x^4, y, z) of $\mathbb{C}\{x, y, z\}$ of a curvilinear quadruple point of V, let :

$$
\begin{aligned}
F_1(x, y, z, \underline{a}) &= x^4 + a_1 x^3 + a_2 x^2 + a_3 x + a_4 \\
F_2(x, y, z, \underline{a}) &= y + a_5 x^3 + a_6 x^2 + a_7 x + a_8 \\
F_3(x, y, z, \underline{a}) &= z + a_9 x^3 + a_{10} x^2 + a_{11} x + a_{12} \\
\underline{a} &= (a_1, \ldots, a_{12}) \quad.
\end{aligned}
$$

For each element f in $\mathcal{O}_P\{x, y, z\}$, the remainder h of the division of f by F_1, F_2, F_3 is of the form :

$$h = h_0(\underline{a}) + h_1(\underline{a})x + h_2(\underline{a})x^2 + h_3(\underline{a})x^3 \quad,$$

where $h_0(\underline{a}), h_1(\underline{a}), h_2(\underline{a})$ and $h_3(\underline{a})$ are in $\mathcal{O}_P$.

$[\mathcal{R}es_5]$ if (f_1, f_2, f_3) is the ideal (x^2, y^2, z) of $\mathbb{C}\{x, y, z\}$ of a square quadruplet of V, let :

$$
\begin{aligned}
F_1(x, y, z, \underline{a}) &= x^2 + a_1 xy + a_2 x + a_3 y + a_4 \\
F_2(x, y, z, \underline{a}) &= y^2 + a_5 xy + a_6 x + a_7 y + a_8 \\
F_3(x, y, z, \underline{a}) &= z + a_9 xy + a_{10} x + a_{11} y + a_{12} \\
\underline{a} &= (a_1, \ldots, a_{12}) \quad .
\end{aligned}
$$

For each element f in $\mathcal{O}_P$, the remainder h of the division of f by F_1, F_2, F_3 is in this case of the form :

$$
h = h_0(\underline{a}) + h_1(\underline{a})x + h_2(\underline{a})xy + h_3(\underline{a})y \quad ,
$$

where $h_0(\underline{a}), h_1(\underline{a}), h_2(\underline{a})$ and $h_3(\underline{a})$ are in $\mathcal{O}_P$.

$[\mathcal{R}es_6]$ If (f_1, f_2, f_3, f_4) is the ideal (x^3, xy, y^2, z) of an elongated quadruplet of V, let :

$$
\begin{aligned}
F_1(x, y, z, \underline{a}) &= x^3 + a_1 x^2 + a_2 x + a_3 y + a_4 \\
F_2(x, y, z, \underline{a}) &= xy + a_5 x^2 + a_6 x + a_7 y + a_8 \\
F_3(x, y, z, \underline{a}) &= y^2 + a_9 x^2 + a_{10} x + a_{11} y + a_{12} \\
F_4(x, y, z, \underline{a}) &= z + a_{13} x^2 + a_{14} x + a_{15} y + a_{16} \\
\underline{a} &= (a_1, \ldots, a_{16}) \quad .
\end{aligned}
$$

For each element f in $\mathcal{O}_P\{x, y, z\}$, the remainder h of the division of f by F_1, F_2, F_3, F_4 is of the form :

$$
h = h_0(\underline{a}) + h_1(\underline{a})x + h_2(\underline{a})x^2 + h_3(\underline{a})y \quad ,
$$

where $h_0(\underline{a}), h_1(\underline{a}), h_2(\underline{a})$ and $h_3(\underline{a})$ are in $\mathcal{O}_P$.

$[\mathcal{R}es_7]$ If $(f_1, \ldots, f_6)$ is the ideal $(x^2, xy, xz, y^2, yz, z^2)$ of a spherical quadruplet of V, let :

$$
\begin{aligned}
F_1(x, y, z, \underline{a}) &= x^2 + a_1 x + a_2 y + a_3 z + a_4 \\
F_2(x, y, z, \underline{a}) &= xy + a_5 x + a_6 y + a_7 z + a_8 \\
F_3(x, y, z, \underline{a}) &= xz + a_9 x + a_{10} y + a_{11} z + a_{12} \\
F_4(x, y, z, \underline{a}) &= y^2 + a_{13} x + a_{14} y + a_{15} z + a_{16} \\
F_5(x, y, z, \underline{a}) &= yz + a_{17} x + a_{18} y + a_{19} z + a_{20} \\
F_6(x, y, z, \underline{a}) &= z^2 + a_{21} x + a_{22} y + a_{23} z + a_{24} \\
\underline{a} &= (a_1, \ldots, a_{24}) \quad .
\end{aligned}
$$

For each element f in $\mathcal{O}_P\{x, y, z\}$, the remainder h of the division of f by $F_1, \ldots, F_6$ is of the form :

$$
h = h_0(\underline{a}) + h_1(\underline{a})x + h_2(\underline{a})y + h_3(\underline{a})z \quad ,
$$

where $h_0(\underline{a}), h_1(\underline{a}), h_2(\underline{a})$ and $h_3(\underline{a})$ are in $\mathcal{O}_P$.

In each case, the module $\dfrac{\mathcal{O}_P\{x, y, z\}}{(F_i)}$ is $\mathcal{O}_P$-flat by definition. Consequently, an element f in $\mathcal{O}_P\{x, y, z\}$ belongs to (F_i) if and only if its remainder is zero.

Remark 8 : The above result is an application of the division theorem of a family with parameters of Galligo ([Gal] 1.2.7 p. 126) in the following cases :

(i) (f_1, f_2, f_3) is a standard basis of the ideal I and the quotient $\dfrac{\mathcal{O}_P\{x, y, z\}}{(x^2, y, z)}$ is a $\mathbb{C}$-vector space of basis $\{\overline{1}, \overline{x}\}$.

(ii) (f_1, f_2, f_3) is a standard basis of the ideal I and the quotient $\dfrac{\mathcal{O}_P\{x, y, z\}}{(x^3, y, z)}$ is a $\mathbb{C}$-vector space of basis $\{\overline{1}, \overline{x}, \overline{x^2}\}$.

(iii) (f_1, f_2, f_3, f_4) is a standard basis of the ideal I and the quotient $\dfrac{\mathcal{O}_P\{x, y, z\}}{(x^2, xy, y^2, z)}$ is a $\mathbb{C}$-vector space of basis $\{\overline{1}, \overline{x}, \overline{y}\}$.

(iv) (f_1, f_2, f_3) is a standard basis of the ideal I and the quotient $\dfrac{\mathcal{O}_P\{x, y, z\}}{(x^4, y, z)}$ is a $\mathbb{C}$-vector space of basis $\{\overline{1}, \overline{x}, \overline{x^2}, \overline{x^3}\}$.

(v) (f_1, f_2, f_3) is a standard basis of the ideal I and the quotient $\dfrac{\mathcal{O}_P\{x, y, z\}}{(x^2, y^2, z)}$ is a $\mathbb{C}$-vector space of basis $\{\overline{1}, \overline{x}, \overline{xy}, \overline{y}\}$.

(vi) (f_1, f_2, f_3, f_4) is a standard basis of the ideal I and the quotient $\dfrac{\mathcal{O}_P\{x, y, z\}}{(x^3, xy, y^2, z)}$ is a $\mathbb{C}$-vector space of basis $\{\overline{1}, \overline{x}, \overline{x^2}, \overline{y}\}$.

(vii) $(f_1, \ldots, f_6)$ is a standard basis of the ideal I and the quotient $\dfrac{\mathcal{O}_P\{x, y, z\}}{(x, y, z)^2}$ is a $\mathbb{C}$-vector space of basis $\{\overline{1}, \overline{x}, \overline{y}, \overline{z}\}$.

c) Finally, note that in our study divisions must be performed by ideals in a polynomials ring. There exists a division algorithm for this purpose (see [E] for example), which was implemented in the algebraic geometry package **Macaulay**. This package was used to perform most of these divisions.

Remark 9 : Let Γ_f be the graph of a morphism $f : X \to Y$ of varieties. The variety $\Gamma_f \subset X \times Y$ is a subvariety of $X \times Y$, isomorphic to X via the first projection. If $(x_1, \ldots, x_n, y_1, \ldots, y_m)$ denotes a local chart of the product $\mathbb{C}^n \times \mathbb{C}^m$ at $(0,0)$, we will say improperly that *the parameters $y_1, \ldots, y_m$ are obtained locally as graphs in terms of the other coordinates $x_1, \ldots, x_n$*, if $y_1, \ldots, y_m$ are elements of the algebra $\mathbb{C}\{x_1, \ldots, x_n\}$. In this case, let $y_i = f_i(x_1, \ldots, x_n)$; the subvariety of $\mathbb{C}^n \times \mathbb{C}^m$ defined by $\{y_i - f_i(x_1, \ldots, x_n) = 0\}_{1 \leq i \leq m}$ is locally isomorphic to the variety $\mathbb{C}^n$.

6.4.1 Non-singularity of $B(V)$ at $\tilde{q}_o$ where q_o is a locally complete intersection quadruple point

Recall that the rational application *residual Res* $: I(V) \cdots \to H^2(V)$ maps (d, q) on to $d' = q \setminus d$ and that $B(V)$ denotes the closure of its graph. Recall also that the projection from $B(V)$ onto $H^4(V)$ is denoted by π.

The case of the curvilinear quadruple point

We intend to prove the following proposition :

Proposition 5 *The variety $B(V)$ is smooth at $\bar{q}_o$ where $q_o \in H^4(V)$ is a curvilinear quadruple point.*

Proof :

We are going to provide a chart of $I(V)$ in the neighborhood of (d_o, q_o) and to express the residual map Res in these coordinates.

The curvilinear quadruplet q_o of support p is defined by the ideal $I(q_o) = (x^4, y, z)$ of $\mathcal{O}_V$. A quadruplet q close to q_o in $H^4(V)$ is given by the ideal

$$I(q) = (x^4 + a_1 x^3 + a_2 x^2 + a_3 x + a_4, \quad y + a_5 x^3 + a_6 x^2 + a_7 x + a_8,$$
$$z + a_9 x^3 + a_{10} x^2 + a_{11} x + a_{12}) \quad ,$$

so that $\underline{a} = (a_1, \ldots, a_{12})$ constitutes a chart of $H^4(V)$ at q_o.

Such a curvilinear quadruple point, i.e. subscheme of a non-singular curve $\mathcal{C}$, can contain only one doublet : it is the doublet with the same support the point p, subscheme of $\mathcal{C}$. Therefore, the doublet d_o is defined by the ideal (x^2, y, z). A doublet d in a neighborhood of d_o in $H^2(V)$ is given by the ideal

$$I(d) = (x^2 + ax + b, -y + cx + d, -z + ex + f) \quad ,$$

so that $\mathcal{A} = (a, b, c, d, e, f)$ constitutes a chart of $H^2(V)$ at d_o. (The minus signs are just for convenience.)

The inclusion of schemes $d \subset q$ in the neighborhood of (d_o, q_o) is equivalent to the inclusion of ideals $I(q) \subset I(d)$ of $\mathbb{C}\{\underline{a}, \mathcal{A}\}\{x, y, z\}$, which can be formulated by equations in $\mathbb{C}\{\underline{a}, \mathcal{A}\}$ that we are going to determine. For each of the three generators of the ideal $I(q)$, the division by the ideal $I(d)$ is performed. The remainder of the division of the ith generator of $I(q)$ is denoted by r_i. According to result $[\mathcal{R}es_1]$, § 6.4.0, the remainders r_i are of the form :

$$r_i = r_{i,0} + r_{i,1} x$$

where $r_{i,0}$ and $r_{i,1}$ are elements of $\mathbb{C}\{\underline{a}, \mathcal{A}\}$.

From $[\mathcal{R}es_1]$ again, the ideal $I(q)$ is contained in the ideal $I(d)$ if and only if the three remainders r_1, r_2, r_3 are identically zero, which can be expressed by the six equations in $\mathbb{C}[\underline{a}, \mathcal{A}]$:

$$a_1 a^2 - a_1 b - a_2 a + a_3 - a^3 + 2ab = 0 \tag{6.1}$$

$$a_1 ab - a_2 b + a_4 - a^2 b + b^2 = 0 \tag{6.2}$$

$$a_5 a^2 - a_5 b - a_6 a + a_7 + c = 0 \tag{6.3}$$

$$a_5 ab - a_6 b + a_8 + d = 0 \tag{6.4}$$

$$a_9 a^2 - a_9 b - a_{10} a + a_{11} + e = 0 \tag{6.5}$$

$$a_9 ab - a_{10} b + a_{12} + f = 0 \tag{6.6}$$

These six equations form the generators of the ideal of $\mathbb{C}\{\underline{a}, \underline{A}\}$ which defines $I(V)$ in $H^2(V) \times H^4(V)$ locally at (d_o, q_o). From these six equations, one can see that the variety $I(V)$ is obtained locally as the graph of a morphism :

$$\mathbb{C}^{12} \to \mathbb{C}^6$$

$$\underline{C} = (a_1, a_2, a_5, a_6, a_9, a_{10}, a, b, c, d, e, f) \mapsto (a_3, a_4, a_7, a_8, a_{11}, a_{12})$$

So $\underline{C}$ forms local coordinates of $I(V)$ at (d_o, q_o).

Let us see now how the residual map Res can be expressed in these coordinates.

As the quadruplet q_o contains only one doublet, the residual doublet $d'_o = q_o \setminus d_o$ is defined by the ideal $I(d'_o) = (x^2, y, z)$ of $\mathcal{O}_V$. A doublet d' close to d'_o in $H^2(V)$ is given by the ideal

$$I(d') = (x^2 + a'x + b', -y + c'x + d', -z + e'x + f') \quad .$$

Thus, the parameters $\underline{A}' = (a', b', c', d', e', f')$ form a chart of $H^2(V)$ at d'_o. The scheme-theoretic inclusion $d' \subset q$ of the residual doublet is equivalent to the inclusion of ideals $I(q) \subset I(d')$ and it is expressed by the six equations $(6.1)'$–$(6.6)'$ of $\mathbb{C}[\underline{a}, \underline{A}']$ (where $a, \ldots, f$ have been replaced by $a', \ldots, f'$ in the equations (6.1)–(6.6)). From $(6.3)'$ and the expression of a_7 (given by (6.3)), it follows that :

$$c' = a_5(a^2 - a'^2) + a_5(b' - b) + a_6(a' - a) + c \tag{6.7}$$

From $(6.4)'$ and the expression of a_8 (given by (6.4)), it follows that :

$$d' = a_5(ab - a'b') + a_6(b' - b) + d \tag{6.8}$$

Equation $(6.5)'$ and the expression of a_{11} (given by (6.5)) yield :

$$e' = a_9(a^2 - a'^2) + a_9(b' - b) + a_{10}(a' - a) + e \tag{6.9}$$

From $(6.6)'$ and the expression of a_{12} (given by (6.6)), we get the equality :

$$f' = a_9(ab - a'b') + a_{10}(b' - b) + f \tag{6.10}$$

Let us now find the ideal of $\mathbb{C}\{\underline{a}, \underline{A}, \underline{A}'\}$ which expresses the inclusion of ideals $I(d) \cdot I(d') \subset I(q)$ in the neighborhood of (d_o, q_o, d'_o). For each of the nine generators of the product $I(d) \cdot I(d')$, one performs the division by the ideal $I(q)$. If R_i denotes the remainder of the division of the ith generator of $I(d) \cdot I(d')$, one has from $[\mathcal{R}es_4]$:

$$R_i = R_{i,0} + R_{i,1}x + R_{i,2}x^2 + R_{i,3}x^3$$

where $R_{i,0}, R_{i,1}, R_{i,2}$ and $R_{i,3}$ are elements of $\mathbb{C}\{\underline{a}, \mathcal{A}, \mathcal{A}'\}$.

Again from $[\mathcal{R}es_4]$, the ideal $I(d) \cdot I(d')$ is contained in $I(q)$ if and only if the nine remainders R_i are identically zero, which yields the $9 \times 4 = 36$ equations :

$$\{R_{ij} = 0\}_{\substack{1 \le i \le 9 \\ 0 \le j \le 3}} \quad .$$

If the remainder of the division of the element $(x^2 + ax + b)(x^2 + a'x + b')$ is denoted by R_1, one has in particular the expressions :

$$\begin{aligned}
R_{1,3} &= a + a' - a_1 \\
R_{1,2} &= b + b' + aa' - a_2 \quad .
\end{aligned}$$

The expression of the coordinate a' of the doublet d' in the chart $\underline{\mathcal{C}}$ of $I(V)$ is obtained by the equation $R_{1,3} = 0$:

$$a' = a_1 - a \tag{6.11}$$

From the expression of a', the equation $R_{1,2} = 0$ can be rewritten :

$$b' = -b + a(a - a_1) + a_2 \quad . \tag{6.12}$$

a' and b' are then replaced by their expressions in terms of a, b, a_1, a_2 in the equations (6.7), (6.8), (6.9) and (6.10). New equations are obtained :

$$\begin{aligned}
c' &= a_5(a^2 + a_1a - a_1^2 - 2b + a_2) + a_6(a_1 - 2a) + c \tag{6.13} \\
d' &= a_5(a^3 - 2a^2a_1 + aa_1^2 + a2 + a_1b - a_1a_2) \\
&\quad + a_6(-2b + a(a - a_1) + a_2) + d \tag{6.14} \\
e' &= a_9(a^2 + a_1a - a_1^2 - 2b + a_2) + a_{10}(a_1 - 2a) + e \tag{6.15} \\
f' &= a_9(a^3 - 2a^2a_1 + aa_1^2 + a2 + a_1b - a_1a_2) \\
&\quad + a_{10}(-2b + a(a - a_1) + a_2) + f \tag{6.16}
\end{aligned}$$

From these six equations, the coordinates of the residual doublet d' can be expressed in the chart $\underline{\mathcal{C}}$ of $I(V)$. Then we check with **Macaulay** that when $a', \ldots, f'$ are replaced by their expressions in terms of $(a_1, a_2, a_5, a_6, a_9, a_{10}, a, b, c, d, e, f)$, the equations (6.1)', (6.2)', $\{R_{ij} = 0\}_{\substack{1 \le i \le 9 \\ 0 \le j \le 3}}$ are satisfied. Thus, the variety $B(V)$ is locally obtained as the graph of a morphism :

$$\mathbb{C}^{12} \quad \rightarrow \quad \mathbb{C}^{12}$$

$$\underline{\mathcal{C}} = (a_1, a_2, a_5, a_6, a_9, a_{10}, a, b, c, d, e, f) \mapsto (a_3, a_4, a_7, a_8, a_{11}, a_{12}, a', b', c', d', e', f') \tag{6.17}$$

So, the variety $B(V)$ is smooth at $\tilde{q}_o$, since it is locally isomorphic to the variety $\mathbb{C}^{12}$. This proves proposition 5.

The case of the square quadruplet

Now, we are going to prove the non-singularity of the variety $B(V)$ at $\tilde{q}$ where $q \in H^4(V)$ is a square quadruplet, which is represented by the following symbol :

First, we state a lemma which gives the form of the elements $\tilde{q}$ in the fiber over such a quadruplet :

Lemma 10 *Let q_o be a square quadruplet of support the point p of V. The elements of the fiber $\pi^{-1}(q_o) \subset B(V)$ over the quadruplet q_o are of the form $(d_\alpha, q_o, d_{-\alpha})_{\alpha \in \mathbb{C}}$ where d_α is the doublet contained in the quadruplet q_o of direction α.*

Proof :

The quadruplet q_o is defined by the ideal (x^2, y^2, z) of $\mathcal{O}_V$ (cf. definition 7.(ii)). A doublet in the quadruplet q_o is fixed. Up to a change of coordinates from x to y, one can always assume that this doublet is defined by the ideal $(x^2, y + \alpha x, z)$, where α is a scalar. The figure below represents the element (d_α, q_o) of $I(V)$:

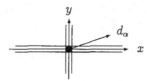

The residual doublet is necessarily contained in the open set of the non-vertical doublets. Indeed, the ideal correponding to the vertical doublet is $I = (y^2, x, z)$ and the inclusion $I(d_\alpha) \cdot I \subset I(q_o)$ is not satisfied because the element $(y + \alpha x)x$ does not belong to the ideal $I(q_o) = (x^2, y^2, z)$. Thus, the residual doublet d' is defined by an ideal of the form $(x^2, y + \beta x, z)$. The inclusion $I(d_\alpha) \cdot I(d') \subset I(q_o)$ will be verified if and only if the element $(y + \alpha x)(y + \beta x)$ is in $I(q_o) = (x^2, y^2, z)$, which is equivalent to the condition $\alpha + \beta = 0$. $\qquad \square$

Let us now prove the following proposition :

Proposition 6 *The variety $B(V)$ is non-singular at each element $(d, q_o, d') \in B(V)$ where $q_o \in H^4(V)$ is a square quadruplet.*

Proof :

In order to prove the non-singularity of the variety $B(V)$ at each element (d, q_o, d') where q_o is a square quadruplet, we will just prove the non-singularity at the most degenerate element, i.e. when the two doublets d and d' are identical and horizontal (cf. lemma 10). The different elements of the fiber $\pi^{-1}(q_o) \subset B(V)$ are represented below :

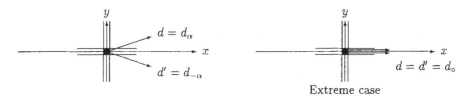

Extreme case

Computations performed for the extreme case will yield the non-singularity of the variety $B(V)$ at each element (d, q_o, d').

As the square quadruplet q_o is defined by the ideal (x^2, y^2, z), a quadruplet q close to q_o in $H^4(V)$ is given by the ideal

$$I(q) = (x^2 + a_1 xy + a_2 x + a_3 y + a_4, \quad y^2 + a_5 xy + a_6 x + a_7 y + a_8,$$
$$z + a_9 xy + a_{10} x + a_{11} y + a_{12}) ,$$

so that $\underline{a} = (a_1, \dots, a_{12})$ forms a chart of $H^4(V)$ at q_o. At the most degenerate point $\tilde{q}_o = (d_o, q_o, d'_o)$ of the fiber $\pi^{-1}(q_o) \subset B(V)$, the two doublets are identical and defined by the same ideal (x^2, y, z) (cf. lemma 10). As before, a doublet d in the neighborhood of d_o in $H^2(V)$ is given by the ideal

$$I(d) = (x^2 + ax + b, -y + cx + d, -z + ex + f) \quad . \tag{6.18}$$

The chart of $H^2(V)$ at d_o is denoted by $\underline{A} = (a, b, c, d, e, f)$. Similarly, the ideal $(x^2 + a'x + b', -y + c'x + d', -z + e'x + f')$ defines a doublet d' close to d'_o. And the chart of $H^2(V)$ at d'_o is denoted by $\underline{A}' = (a', b', c', d', e', f')$.

Let us determine the local equations of the subvariety $I(V)$ of $H^2(V) \times H^4(V)$ at (d_o, q_o). We have to determine the ideal of $\mathbb{C}\{\underline{a}, \underline{A}\}$ which expresses the inclusion $d \subset q$ in the scheme-theoretic sense. To do so, one performs the division of each generator of $I(q)$ by the ideal $I(d)$. If l_i denotes the remainder of the division of the ith generator of $I(q)$, one has from $[\mathcal{R}es_1]$:

$$l_i = l_{i,0} + l_{i,1} x$$

where $l_{i,0}$ and $l_{i,1}$ are elements of $\mathbb{C}\{\underline{a}, \underline{A}\}$. From $[\mathcal{R}es_1]$ again, the inclusion $I(q) \subset I(d)$ will be satisfied if and only if the remainders l_i are non zero, which can be expressed by the six equations $\{l_{i,j} = 0\}_{\substack{i = 1, 2, 3 \\ j = 0, 1}}$ in $\mathbb{C}[\underline{a}, \underline{A}]$. More explicitly :

$$l_{1,1} = -a_1 ac + a_1 d + a_2 + a_3 c - a \tag{6.19}$$

$$l_{1,0} = -a_1bc + a_3d + a_4 - b \tag{6.20}$$

$$l_{2,1} = -a_5ac + a_5d + a_6 + a_7c - ac^2 + 2cd \tag{6.21}$$

$$l_{2,0} = -a_5bc + a_7d + a_8 - bc^2 + d^2 \tag{6.22}$$

$$l_{3,1} = -a_9ac + a_9d + a_{10} + a_{11}c + e \tag{6.23}$$

$$l_{3,0} = -a_9bc + a_{11}d + a_{12} + f \tag{6.24}$$

Therefore these six equations define the subvariety $I(V)$ of $H^2(V) \times H^4(V)$ locally at (d_o, q_o). These six equations (6.19)–(6.24), taken in this order, enable one to express $a_2, a_4, a_6, a_8, a_{10}$ and a_{12} as functions of the other coordinates. Thus, $\mathcal{C} = (a_1, a_3, a_5, a_7, a_9, a_{11}, a, b, c, d, e, f)$ forms a chart of $I(V)$ at (d_o, q_o).

Let us now determine the residual application Res in these coordinates. The inclusion of the residual doublet d' in the quadruplet q in the neighborhood of (q_o, d'_o) is expressed by the six equations $\{l'_{i,j} = 0\}_{\substack{i=1,2,3 \\ j=0,1}}$ in $\mathbb{C}[\underline{a}, \underline{A}']$ where $l'_{i,j}$ is obtained from $l_{i,j}$ by replacing $a, \ldots, f$ by $a', \ldots, f'$. In particular, we get from the inclusion of ideals $I(d) \cdot I(d') \subset I(q)$ that the element $(-y + cx + d)(y - c'x - d')$ is in the ideal $I(q)$. From $[\mathcal{R}es_5]$, the remainder of the division of this element by $I(q)$ is of the form $r_0 + r_1x + r_2xy + r_3y$, where r_0, r_1, r_2 and r_3 are in $\mathbb{C}\{\underline{a}, \underline{A}', \underline{A}\}$, and the condition $(-y + cx + d)(y - c'x - d') \in I(q)$ can be expressed by the four equations $\{r_i = 0\}_{i=0,\ldots,3}$. In particular, we get the expressions :

$$r_2 = c + c' + a_1cc' + a_5$$
$$r_3 = d + d' + a_7 + a_3cc' \quad .$$

The equation $r_2 = 0$, which can be rewritten as $c'(1 + a_1c) = -a_5 - c$, enables one to express the parameter c' locally as a graph because the element $(1 + a_1c)$ is invertible in the ring $\mathbb{C}\{\mathcal{C}\}$. Then one obtains the expression :

$$c' = -(a_5 + c)(1 + a_1c)^{-1} \quad . \tag{6.25}$$

The equation $r_3 = 0$ and the expression of c' yield the following expression for the parameter d' in $\mathbb{C}\{\mathcal{C}\}$:

$$d' = -d - a_7 + a_3c(a_5 + c)(1 + a_1c)^{-1} \tag{6.26}$$

From the expression of a_2 (given by (6.19)) and the equation $l'_{1,1} = 0$, one gets the equality :

$$a'(1 + a_1c') = a(1 + a_1c) + a_1(d' - d) + a_3(c' - c)$$

From the expression of a_4 (given by (6.20)) and the equation $l'_{1,0} = 0$, one gets the equality :

$$b'(1 + a_1c') = b(1 + a_1c) + a_3(d' - d)$$

As the element $1 + a_1 c'$ ($= 1 - a_1(a_5 + c)(1 + a_1 c)^{-1}$, by (6.25)) is invertible in the ring $\mathbb{C}\{\underline{\mathcal{C}}\}$, one rewrites the two following equalities :

$$a' = [a(1 + a_1 c) + a_1(d' - d) + a_3(c' - c)](1 + a_1 c')^{-1} \qquad (6.27)$$
$$b' = [b(1 + a_1 c) + a_3(d' - d)](1 + a_1 c')^{-1} \qquad (6.28)$$

c' and d' are then replaced by their expressions in terms of a_1, a_3, a_5, a_7, c and d in the equations (6.27) and (6.28), which enables us to express the coordinates a' and b' of the residual doublet in the chart $\underline{\mathcal{C}}$. From the expression of a_{10} (given by (6.23)) and the equation $l'_{3,1} = 0$, one gets :

$$e' = e - a_{11}(c' - c) + a_9(d - d') + a_9 a' c' - a_9 a c \qquad (6.29)$$

The expression of a_{12} (given by (6.24)) and the equation $l'_{3,0} = 0$ yield :

$$f' = f + a_9(b'c' - bc) + a_{11}(d - d') \qquad (6.30)$$

It was shown above that a', b', c' and d' are elements of the algebra $\mathbb{C}\{\underline{\mathcal{C}}\}$. When a', b', c' and d' are replaced by their expressions in terms of the coordinates $\underline{\mathcal{C}}$ in the equations (6.29) and (6.30), one gets the expressions of the last two coordinates e' and f' of the residual doublet in the chart $\underline{\mathcal{C}}$. The coordinates $a', \ldots, f'$ of the residual doublet have thus been obtained locally as graphs in the variety $\mathbb{C}^{12}$ of coordinates $\underline{\mathcal{C}}$.

Then one checks that when $a', \ldots, f'$ are replaced by their expressions in terms of the coordinates $\underline{\mathcal{C}}$ in the equations $l'_{2,0} = 0$, $l'_{2,1} = 0$ and in the equations given by the generators of the ideal of $\mathbb{C}\{\underline{a}, \underline{A}, \underline{A'}\}$ which express the inclusion $I(d) \cdot I(d') \subset I(q)$, one gets indeed $0 = 0$. Therefore, the variety $B(V)$ is locally isomorphic at $\tilde{q}_o$ to the graph of a morphism :

$$\mathbb{C}^{12} \rightarrow \mathbb{C}^{12}$$
$$(a_1, a_3, a_5, a_7, a_9, a_{11}, a, b, c, d, e, f) \mapsto (a_2, a_4, a_6, a_8, a_{10}, a_{12}, a', b', c', d', e', f')$$

The variety $B(V)$ is smooth at this point and $(a_1, a_3, a_5, a_7, a_9, a_{11}, a, b, c, d, e, f)$ forms a chart.

Remark 10 One can also choose $(a_1, a_3, a_9, a_{11}, a, b, c, d, e, f, c', d')$ as chart of $B(V)$ at $\tilde{q}_o$. In this case, the variety $B(V)$ is locally isomorphic to the graph of a morphism from $\mathbb{C}^{12}$ to $\mathbb{C}^{12}$ which is expressed by the following equations :

$$
\begin{aligned}
a_2 &= a(1 + a_1c) - a_1d - a_3c \\
a_4 &= b(1 + a_1c) - a_3d \\
a_5 &= -(c + c' + a_1cc') \\
a_6 &= -acc'(1 + a_1c) + dc'(1 + a_1c) + c(d' + a_3cc') \\
a_7 &= -(d + d' + a_3cc') \\
a_8 &= -bcc'(1 + a_1c) + d(d' + a_3cc') \\
a_{10} &= -a_{11}c - e - a_9d + a_9ac \\
a_{12} &= a_9bc - a_{11}d - f \\
a'(1 + a_1c') &= a(1 + a_1c) + a_1(d' - d) + a_3(c' - c) \\
b'(1 + a_1c') &= b(1 + a_1c) + a_3(d' - d) \\
(e' - e)(1 + a_1c') &= (c' - c)[-a_{11}(1 + a_1c') + a_9a + a_3a_9c'] + a_9(d' - d) \\
(f' - f)(1 + a_1c') &= a_9b(c' - c) + (d' - d)[a_3a_9c' - a_{11}(1 + a_1c')]
\end{aligned}
$$

These charts will be used for the construction of the variety $\widehat{H^4(V)}$ in the product $[\widehat{H^3(V)}]^4 \times [B(V)]^6$ (cf. § 7.1.2).

From lemma 10, the other elements of the fiber $\pi^{-1}(q_o) \subset B(V)$ over the square quadruplet q_o are of the form $(d_\alpha, q_o, d_{-\alpha})_{\alpha \in \mathbb{C}}$, where the doublet d_α is defined by the ideal $(x^2, y + \alpha x, z)$. The computations performed in the neighborhood of $\tilde{q}_o = (d_0, q_o, d_0)$ in order to establish a chart of $B(V)$ are still valid for $(d_\alpha, q_o, d_{-\alpha})$. It is sufficient to replace c by $\mathbf{C} - \alpha$ and c' by $\mathbf{C}' + \alpha$ in the equations, where $\mathbf{C}$ and $\mathbf{C}'$ are local parameters. This ends the proof of proposition 6.

Propositions 5 and 6 lead to the non-singularity of the variety $B(V)$ at each element $\tilde{q}$ where $q \in H^4(V)$ is a locally complete intersection quadruple point. We now proceed in proving the same result when q is no longer a complete intersection.

6.4.2 Non-singularity of $B(V)$ at $\tilde{q}_o$ where q_o is a non-locally complete intersection quadruple point

Recall (cf. definitions 7) that the non-locally complete intersection quadruple points are the elongated and spherical quadruplets. Recall also that $R(V)$ denotes the subvariety of $H^2(V) \times H^4(V) \times H^2(V)$, the elements (d, q, d') of which satisfy the inclusions :

$$
I(d) \cdot I(d') \subset I(q) \subset I(d) \cap I(d') \quad .
$$

The variety $R(V)$ contains the graph Γ_{Res} of the rational residual map $Res : I(V) \cdots \to H^2(V)$ since elements $(d, q, d' = q \setminus d)$ satisfy the previous inclusions. Therefore, the variety $R(V)$ contains $B(V)$, the closure of the graph Γ_{Res} in $I(V) \times H^2(V)$.

The case of the elongated quadruplet

Recall that if $q_o \in H^4(V)$ is the elongated quadruplet of support the point p of V, in an appropriate local coordinate system centered at p, it is defined by the ideal (x^3, xy, y^2, z) (cf. definition 7-(iii)). The following lemma provides a description of the elements (d, q_o, d') of $B(V)$:

Lemma 11 *One of the two doublets d, d' which constitutes the element (d, q_o, d') of $B(V)$ defines the same line as the one defined by the elongated quadruplet q_o.*

Proof :

The two doublets d and d' have the same support as the quadruplet q_o. Therefore, the question is to prove that one of these two doublets is a subscheme of the curve of local equations $y = 0 = z$. In this system of local coordinates (x, y, z), the doublets d and d' are necessarily defined by ideals of the form :

$$
\begin{aligned}
I(d) &= (x^2, y + \lambda x, z) \\
I(d') &= (x^2, y + \lambda' x, z) \quad ,
\end{aligned}
$$

where λ and λ' are elements in $\mathbb{C}$. Then, the scheme-theoretic inclusions $d, d' \subset q_o$ follow. Then one checks that the inclusion $I(d) \cdot I(d') \subset I(q_o)$ is equivalent to the condition

$$
\lambda \lambda' x^2 \in I(q_o) = (x^3, xy, y^2, z)
$$

This condition is equivalent to $\lambda \lambda' = 0$.   □

According to our drawing conventions, the elements (d, q_o, d') in the fiber $\pi^{-1}(q_o) \subset B(V)$ are of the form :

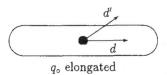

q_o elongated

with as extreme case the element $\tilde{q}_o = (d_0, q_o, d'_0)$ where the doublets d_0 et d'_0 are identical and define the same line as the quadruplet :

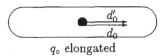

q_o elongated

Let us now prove the non-singularity of $B(V)$ at this element $\tilde{q}_o$. The two doublets d_0 and d'_0 are defined by the ideal (x^2, y, z). As before (see (6.18)),

$\underline{A} = (a, b, c, d, e, f)$ denotes a chart of $H^2(V)$ at d_o and $\underline{A'} = (a', b', c', d', e', f')$ a chart of $H^2(V)$ at d'_o. A quadruplet q close to q_o in $H^4(V)$ is defined by the ideal :

$$I(q) = (\quad x^3 + a_1 x^2 + a_2 x + a_3 y + a_4, \, xy + a_5 x^2 + a_6 x + a_7 y + a_8,$$
$$y^2 + a_9 x^2 + a_{10} x + a_{11} y + a_{12}, \, z + a_{13} x^2 + a_{14} x + a_{15} y + a_{16})$$

where a_4, a_8, a_{10} and a_{12} are explicit functions of $a_1, a_2, a_3, a_5, a_6, a_7, a_9$ and a_{11} (cf. Appendix B.1), so that $\underline{a} = (a_1, a_2, a_3, a_5, a_6, a_7, a_9, a_{11}, a_{13}, a_{14}, a_{15}, a_{16})$ constitutes a chart of $H^4(V)$ at q_o.

The scheme-theoretic inclusion $d \subset q$ in the neighborhood of (d_o, q_o) is equivalent to the inclusion of ideals $\quad(*)\quad I(q) \subset I(d)\quad$ and it is expressed by an ideal of $\mathbb{C}[\underline{a}, \underline{A}]$ which we are going to determine now. For this purpose, one divides each generator of $I(q)$ by the ideal $I(d) = (x^2 + ax + b, -y + cx + d, -z + ex + f)$. From $[\mathcal{R}es_1]$, the remainder r_i of the division of the ith generator is of the form :

$$r_i = r_{i,0} + r_{i,1} x$$

where $r_{i,0}$ and $r_{i,1}$ are elements of $\mathbb{C}\{\underline{a}, \underline{A}\}$. The inclusion $(*)$ can be expressed by the eight equations $\{r_{i,j} = 0\}_{\substack{1 \le i \le 4 \\ j=0,1}}$ in $\mathbb{C}\{\underline{a}, \underline{A}\}$. More explicitly, one has the expressions :

$$
\begin{aligned}
r_{1,1} \;&=\; -a_1 a + a_2 + a_3 c + a^2 - b & (6.31)\\
r_{1,0} \;&=\; 2a_3 a_5 a_7 - a_1 a_7^2 + a_7^3 - a_3 a_6 + a_2 a_7 + a_3 a_{11}\\
&\quad -a_1 b + a_3 d + ab & (6.32)\\
r_{2,1} \;&=\; -a_5 a + a_6 + a_7 c - ac + d & (6.33)\\
r_{2,0} \;&=\; -a_3 a_5^2 - a_5 a_7^2 + a_6 a_7 - a_3 a_9 - a_5 b + a_7 d - bc & (6.34)\\
r_{3,1} \;&=\; a_1 a_5^2 + a_5^2 a_7 - 2a_5 a_6 + a_1 a_9 - a_7 a_9 + a_5 a_{11} - a_9 a + a_{11} c\\
&\quad -ac^2 + 2cd & (6.35)\\
r_{3,0} \;&=\; a_3 a_5^3 - a_1 a_5^2 a_7 + a_2 a_5^2 + 2a_5 a_6 a_7 + a_3 a_5 a_9 - a_1 a_7 a_9 + a_7^2 a_9 - a_5 a_7 a_{11}\\
&\quad -a_6^2 + a_2 a_9 + a_6 a_{11} - a_9 b + a_{11} d - bc^2 + d^2 & (6.36)\\
r_{4,1} \;&=\; -a_{13} a + a_{14} + a_{15} c + e & (6.37)\\
r_{4,0} \;&=\; -a_{13} b + a_{15} d + a_{16} + f & (6.38)
\end{aligned}
$$

Similarly, the incidence relation $d' \subset q$ can be expressed in the neighborhood of (d'_o, q_o) by the eight equations $\{r'_{i,j} = 0\}_{\substack{1 \le i \le 4 \\ j=0,1}}$ in $\mathbb{C}\{\underline{a}, \underline{A'}\}$, where $r'_{i,j}$ is obtained from $r_{i,j}$ by replacing $a, \ldots, f$ by $a', \ldots, f'$. Let us now exhibit the ideal of $\mathbb{C}\{\underline{a}, \underline{A}, \underline{A'}\}$ which expresses the ideal inclusion $\quad I(d) \cdot I(d') \subset I(q)\quad (**)$ in the neighborhood of (d_o, q_o, d'_o). Let :

$$
\begin{aligned}
I(d) \;&=\; (x^2 + ax + b, -y + cx + d, -z + ex + f)\\
&=\; (f_1, f_2, f_3)\\
I(d') \;&=\; (x^2 + a'x + b', -y + c'x + d', -z + e'x + f')\\
&=\; (f'_1, f'_2, f'_3)
\end{aligned}
$$

One adds to the nine generators of the ideal $I(d) \cdot I(d')$ combinations of elements of $I(q)$, so that one obtains nine new generators $P_1, \cdots, P_9$ of the form :

$$P_i = P_{i,0} + P_{i,1}x + P_{i,2}x^2 + P_{i,3}y \quad .$$

The inclusion (**) is equivalent to the new inclusion $(P_1, \cdots, P_9) \subset I(q)$. This last inclusion can be expressed, from $[\mathcal{R}es_6]$, by the $9 \times 4 = 36$ equations $\{P_{i,j} = 0\}_{\substack{1 \leq i \leq 9 \\ 1 \leq j \leq 3}}$ in $\mathbb{C}[\underline{a}, \mathcal{A}, \mathcal{A}']$. If P_1 (resp. P_2, P_4 and P_5) denotes the remainder of the division by $I(q)$ of $f_1 f_1'$ (resp. $f_2' f_1, f_1' f_2$ and $f_2 f_2'$), one has in particular :

$$\begin{align}
P_{1,2} &= a_1^2 + a_3 a_5 - a_1(a + a') - a_2 + aa' + b + b' \tag{6.39}\\
P_{2,3} &= -a_3 a_5 - a_7^2 - a_3 c' + a_7 a - b \tag{6.40}\\
P_{4,3} &= -a_3 a_5 - a_7^2 - a_3 c + a_7 a' - b' \tag{6.41}\\
P_{5,2} &= a_5(c + c') - a_9 + cc' \tag{6.42}\\
P_{5,3} &= a_7(c + c') - a_{11} - d - d' \tag{6.43}
\end{align}$$

The equations $r_{2,1} = 0, r'_{2,1} = 0, r_{4,1} = 0$ and $r'_{4,1} = 0$ enable one to express the parameters d, d', e and e' as functions of the other coordinates. The equations $P_{2,3} = 0, P_{4,3} = 0$ and $P_{5,2} = 0$ enable one to express b, b' and a_9 in terms of a_3, a_5, a_7, a, c, a' and c'. From the equation $P_{1,2} = 0$ and the expressions of b and b', one gets the following expression for a_2 :

$$a_2 \ \dot{=} \ -a_3(a_5 + c + c') + a_1^2 - 2a_7^2 + aa' + (a + a)(a_7 - a_1) \tag{6.44}$$

The equation $P_{5,3} = 0$ and the expressions of d and d' yield the equality :

$$a_{11} \ = \ 2a_6 + a_7(c + c') - a(a_5 + c) - a'(a_5 + c') \tag{6.45}$$

From the expressions of b and d and the equation $r_{4,0} = 0$, one gets the following expression for f :

$$f \ = \ a_{13}[a_7(a - a_7) - a_3(a_5 + c')] - a_{16} - a_{15}[-a_6 + a(a_5 + c) - a_7 c] \tag{6.46}$$

Similarly, from the expressions of b' and d' and the equation $r'_{4,0} = 0$, one gets :

$$f' \ = \ a_{13}[a_7(a' - a_7) - a_3(a_5 + c)] - a_{16} - a_{15}[-a_6 + a'(a_5 + c') - a_7 c'] \tag{6.47}$$

Thus, the equations (6.33), (6.33)', (6.37), (6.37)', (6.38), (6.38)', (6.39)–(6.43) enable one to express the eleven parameters $a_2, a_9, a_{11}, b, d, e, f, b', d', e', f'$ as functions of the other thirteen coordinates. Let $\mathcal{C} = (a_1, a_3, a_5, a_6, a_7, a_{13}, a_{14}, a_{15}, a_{16}, a, c, a', c')$. These eleven parameters are then replaced by their expressions in terms of the coordinates $\mathcal{C}$ in the other forty-one generators of the variety $R(V)$ (of course, we use **Macaulay**). Only four equations are satisfied. We are left with thirty-seven generators with a zero linear part. The ideal of $\mathbb{C}[\mathcal{C}]$ generated by these thirty-seven generators is denoted by J. Therefore the variety $R(V)$ is locally isomorphic at $(d_\circ, q_\circ, d'_\circ)$ to the subvariety $V(J)$ of $\mathbb{C}^{13}$, of coordinates $\mathcal{C}$.

Lemma 12 *The subvariety $V(J)$ of $\mathbb{C}^{13}$ is reducible. This subvariety possesses two smooth irreducible components, one of dimension 8 and the other of dimension 12.*

Proof :

One shows (still by using **Macaulay**) that the ideal J can be rewritten as a product $J = (h) \cdot L$, where $h = a_1 + a_7 - a - a'$ and L is the ideal $(a_3, a_5 + c, a_5 + c', a_1 - a_7 - a, a_1 - a_7 - a')$ of $\mathbb{C}[\underline{C}]$. The hyperplane $V((h))$ and the linear subspace $V(L)$ of dimension 8 intersect transversally. □

Then the variety $R(V)$ is the union of two smooth irreducible components, one of which is of dimension 12. As recalled at the beginning of § 6.4.0, this variety $R(V)$ contains the variety $B(V)$ which is irreducible of dimension 12, from proposition 4. Then one concludes that the variety $B(V)$ is locally isomorphic to the hyperplane $V((h))$. Therefore the variety $B(V)$ is non-singular at the element $\tilde{q}_o$, the most degenerate of the fiber $\pi^{-1}(q_o) \subset B(V)$. In the neighborhood of this point, the variety $B(V)$ is locally isomorphic to the graph of a morphism :

$$\mathbb{C}^{12} \rightarrow \mathbb{C}^{12}$$
$$(a_1, a_3, a_5, a_6, a_7, a_{13}, a_{14}, a_{15}, a_{16}, a, c, c') \mapsto (a_2, a_9, a_{11}, b, d, e, f, a', b', d', e', f')$$

and $(a_1, a_3, a_5, a_6, a_7, a_{13}, a_{14}, a_{15}, a_{16}, a, c, c')$ constitutes a local chart.

The non-singularity of $B(V)$ at any element of the fiber follows immediately. Indeed, from lemma 11, one can always assume the doublet d to be defined by the ideal $(x^2, y + \alpha x, z)$ where α is a fixed non zero scalar and the doublet d' to be defined by the ideal (x^2, y, z) :

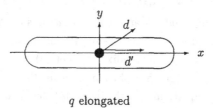

q elongated

A doublet close to d is then defined by the ideal $(x^2 + ax + b, -y + (\mathbf{C} - \alpha)x + d, -z + ex + f)$. c is replaced by $(\mathbf{C} - \alpha)$ everywhere in the computations, where $\mathbf{C}$ is a local parameter. In this case, the variety $R(V)$ is smooth, irreducible of dimension 12 because the subvariety $V(L)$ of $\mathbb{C}^{13}$ is reduced to the empty set – The ideal L of $\mathbb{C}\{a_1, a_3, a_5, a_6, a_7, a_{13}, a_{14}, a_{15}, a_{16}, a, \mathbf{C}, a', c'\}$ contains the element $(a_5 + \mathbf{C} - \alpha)$ which is invertible. From proposition 4, the inclusion $B(V) \subset R(V)$ is in fact an equality. This ends the proof of the following proposition :

Proposition 7 *The variety $B(V)$ is non-singular at every element $(d, q, d') \in B(V)$ where $q \in H^4(V)$ is an elongated quadruplet.*

To complete the proof of the non-singularity of the variety $B(V)$, it remains to study the case of the spherical quadruplet.

The case of the spherical quadruplet

If $q_o \in H^4(V)$ is the spherical quadruplet of support the point p of V, it is defined by the ideal $(x, y, z)^2$ (cf. definition 7.(iv)). In this case, the doublets d and d' which constitute the element (d, q_o, d') of $B(V)$ have the same support the point p and any directions. The extreme configuration is obtained when the doublets are identical. With our drawing conventions, the elements $\tilde{q}_o$ are represented as :

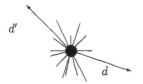

As usual, we prove first the non-singularity of $B(V)$ in the extreme case, i.e. when the two doublets d_o and d'_o have the same direction. One can always assume that these doublets d_o and d'_o are defined by the ideal (x^2, y, z). As usual, $\underline{A} = (a, b, c, d, e, f)$ will denote a local chart of $H^2(V)$ at d_o and $\underline{A}' = (a', b', c', d', e', f')$ a local chart of $H^2(V)$ at d'_o. A quadruplet q close to q_o in $H^4(V)$ is defined by the ideal :

$$I(q) = (\ x^2 + a_1 x + a_2 y + a_3 z + a_4, \ xy + a_5 x + a_6 y + a_7 z + a_8,$$
$$xz + a_9 x + a_{10} y + a_{11} z + a_{12}, \ y^2 + a_{13} x + a_{14} y + a_{15} z + a_{16},$$
$$yz + a_{17} x + a_{18} y + a_{19} z + a_{20}, \ z^2 + a_{21} x + a_{22} y + a_{23} z + a_{24})$$

where $a_4, a_8, a_{12}, a_{16}, a_{20}$ and a_{24} are explicit functions of the other eighteen parameters. These eighteen parameters must also satisfy fifteen linearly independent quadratic equations $\{H_i = 0\}_{1 \leq i \leq 15}$ (cf. appendix 7.3). Let $(\underline{a}) = (a_1, \dots, \tilde{a}_4, \dots, \tilde{a}_8, \dots,$ $\tilde{a}_{12}, \dots, \tilde{a}_{16}, \dots, \tilde{a}_{20}, \dots, \tilde{a}_{24})$ (the notation $\tilde{a}_i$ means that a_i has been removed). Then the fifteen quadratic equations constitute in $\mathbb{C}[\underline{a}]$ the local equations of $H^4(V)$ at q_o. (Recall that the spherical quadruplets are the singular points of the Hilbert scheme $H^4(V)$ (see [I2] , [F]).)

The inclusion of schemes $d \subset q$ in the neighborhood of (d_o, q_o) is equivalent to the inclusion of ideals $I(q) \subset I(d)$ which is expressed by equations in $\mathbb{C}[\underline{a}, \underline{A}]$ that we are going to determine now. To do so, one divides each of the six generators of the ideal $I(q)$ by the ideal $I(d)$. From $[\mathcal{R}es_1]$ again, the remainder r_i of the division of the ith generator is of the form :

$$r_i = r_{i,0} + r_{i,1} x$$

where $r_{i,0}$ and $r_{i,1}$ are elements of $\dfrac{\mathbb{C}\{\underline{a}, \underline{A}\}}{(H_i)_{1 \leq i \leq 15}}$. The previous inclusion of ideals is then equivalent to the inclusion $(r_1, \dots, r_6) \subset I(d)$. However, the ideal $I(d)$ can not contain such non zero affine linear forms $r_i = r_{i,0} + r_{i,1}x$. The previous inclusion is therefore expressed by the $6 \times 2 = 12$ equations $\{r_{i,j} = 0\}_{\substack{1 \le i \le 6 \\ j=0,1}}$. In particular, one has the expressions :

$$r_{1,1} = a_1 + a_2c + a_3e - a \tag{6.48}$$

$$r_{2,1} = a_5 + a_6c + a_7e - ac + d \tag{6.49}$$

$$r_{3,1} = a_9 + a_{10}c + a_{11}e - ae + f \tag{6.50}$$

$$r_{5,1} = a_{17} + a_{18}c + a_{19}e - ace + de + cf \tag{6.51}$$

Similarly, the scheme-theoretic inclusion $d' \subset q$ in the neighborhood of (d'_o, q_o) is expressed by the twelve equations $\{r'_{i,j} = 0\}_{\substack{1 \le i \le 6 \\ j=0,1}}$ where $r'_{i,j}$ is obtained from $r_{i,j}$ by replacing $a, \dots, f$ by $a', \dots, f'$.

Let us now determine the equations in $\mathbb{C}[\underline{a}, \underline{A}, \underline{A'}]$ which express the inclusion of ideals $I(d) \cdot I(d') \subset I(q)$ in the neighborhood of (d_o, q_o, d'_o). One adds to each of the nine generators of the product $I(d) \cdot I(d')$ elements of $I(q)$, in order to obtain nine new generators $R_1, \dots, R_9$ of the form :

$$R_i = R_{i,0} + R_{i,1}x + R_{i,2}y + R_{i,3}z \quad,$$

where $R_{i,j}$ are elements in $\dfrac{\mathbb{C}[\underline{a}, \underline{A}, \underline{A'}]}{(H_i)_{1 \le i \le 15}}$. The inclusion of ideals $I(d) \cdot I(d') \subset I(q)$ is equivalent to the inclusion $(R_1, \dots, R_9) \subset I(q)$. From $[\mathcal{R}es_7]$, the elements R_i are necessarily zero in order to be able to satisfy the previous inclusion, which is expressed by the $9 \times 4 = 36$ equations $\{R_{i,j} = 0\}_{\substack{1 \le i \le 9 \\ 0 \le j \le 3}}$. The remainder of the division by the ideal $I(q)$ of the element $(-y + cx + d)(x^2 + a'x + b')$ (resp. $(x^2 + ax + b)(-y + c'x + d')$, $(-y + cx + d)(-y + c'x + d')$, $(-z + ex + f)(-y + c'x + d)$, $(-y + cx + d)(-z + e'x + f')$ and $(-z + ex + f)(-z + e'x + f'))$ is denoted by R_2 (resp. R_4, R_5, R_6, R_8 and R_9). In particular, one gets the expressions :

$$R_{2,2} = a_1a_2c - a_2a_5 + a_2a_6c - a_6^2 + a_3a_{10}c - a_7a_{10} - a_2ca'$$
$$-a_2d + a_6a' - b' \tag{6.52}$$

$$R_{4,2} = a_1a_2c' - a_2a_5 + a_2a_6c' - a_6^2 + a_3a_{10}c' - a_7a_{10} - a_2c'a$$
$$-a_2d' + a_6a - b \tag{6.53}$$

$$R_{5,1} = -a_1cc' + a_5(c + c') - a_{13} + cd' + dc' \tag{6.54}$$

$$R_{5,2} = -a_2cc' + a_6(c + c') - a_{14} - d - d' \tag{6.55}$$

$$R_{5,3} = -a_3cc' + a_7(c + c') - a_{15} \tag{6.56}$$

$$R_{6,2} = -a_2ec' + a_6e + a_{10}c' - a_{18} - f \tag{6.57}$$

$$R_{8,2} = -a_2ce' + a_6e' + a_{10}c - a_{18} - f' \tag{6.58}$$

$$R_{8,3} = -a_3ce' + a_7e' + a_{11}c - a_{19} - d \tag{6.59}$$

$$R_{9,1} \;=\; -a_1 e e' + a_9(e + e') - a_{21} + e f' + f e' \tag{6.60}$$

$$R_{9,2} \;=\; -a_2 e e' + a_{10}(e + e') - a_{22} \tag{6.61}$$

$$R_{9,3} \;=\; -a_3 e e' + a_{11}(e + e') - a_{23} - f - f' \tag{6.62}$$

Claim : All these equations enable us to express the seventeen parameters $a, a', b, b', d,$ $d', f, f', a_{13}, a_{14}, a_{15}, a_{17}, a_{18}, a_{19}, a_{21}, a_{22}$ and a_{23} in terms of the thirteen other coordinates $a_1, a_2, a_3, a_5, a_6, a_7, a_9, a_{10}, a_{11}, c, e, c', e'$.

Let us prove this claim. Let $\underline{C} = (a_1, a_2, a_3, a_5, a_6, a_7, a_9, a_{10}, a_{11}, c, e, c', e')$. The equations $r_{1,1} = 0$ and $r'_{1,1} = 0$ enable one to express a and a' in terms of the coordinates $\underline{C}$. The equation $r_{2,1} = 0$ and the expression of a yield the expression for d :

$$d \;=\; -a_5 + c(a_1 + a_2 c + a_3 e - a_6) - a_7 e \tag{6.63}$$

Similarly, one obtains the expression for d' :

$$d' \;=\; -a_5 + c'(a_1 + a_2 c' + a_3 e' - a_6) - a_7 e' \tag{6.64}$$

From the equation $r_{3,1} = 0$ and the expression of a, one gets the equality :

$$f \;=\; -a_9 + e(a_1 + a_2 c + a_3 e - a_{11}) - a_{10} c \tag{6.65}$$

The expression for f' :

$$f' \;=\; -a_9 + e'(a_1 + a_2 c' + a_3 e' - a_{11}) - a_{10} c' \tag{6.66}$$

is obtained similarly from the equation $r'_{3,1} = 0$ and the expression of a'.

The equations $R_{5,3} = 0$ and $R_{9,2} = 0$ enable one to obtain a_{15} and a_{22} in terms of the coordinates $\underline{C}$. When d and d' are replaced by their expressions in the equation $R_{5,2} = 0$, one gets the expression for a_{14} :

$$\begin{aligned}
a_{14} \;=\;\; & 2a_5 + (c + c')(2a_6 - a_1) - a_2(c^2 + c'^2 + cc') - a_3(ec + e'c') \\
& + a_7(e + e')
\end{aligned} \tag{6.67}$$

From the equations $R_{5,1} = 0$, (6.63) and (6.64), one gets :

$$a_{13} \;=\; cc'[a_1 + a_2(c + c') + a_3(e + e') - 2a_6] - a_7(ec' + e'c) \tag{6.68}$$

The equations $R_{8,3} = 0$ and (6.63) lead to the expression for a_{19} :

$$a_{19} \;=\; a_5 - c[a_1 + a_2 c + a_3(e + e') - a_6 - a_{11}] + a_7(e + e') \tag{6.69}$$

The equations $R_{6,2} = 0$ and (6.65) yield the expression for a_{18} :

$$a_{18} \;=\; a_9 - e[a_1 + a_2(c + c') + a_3 e - a_6 - a_{11}] + a_{10}(c + c') \tag{6.70}$$

From the equations $r_{5,1} = 0$, (6.63), (6.65) and from the expressions of a_{18} and of a_{19}, one obtains the equality :

$$a_{17} = ec[a_1 - a_6 - a_{11} + a_2(c + c') + a_3(e + e')] - a_{10}cc' - a_7ee' \quad (6.71)$$

From the equations $R_{2,2} = 0$, (6.63) and the expression of a', one obtains :

$$b' = a_6[a_1 + a_2(2c + c') + a_3e' - a_6] - a_2c[a_1 + a_2(c + c') + a_3(e + e')]$$
$$+ a_2a_7e + a_3a_{10}c - a_7a_{10} \quad (6.72)$$

Similarly, the equations $R_{4,2} = 0$, (6.64) and the expression for a yield the expression :

$$b = a_6[a_1 + a_2(c + 2c') + a_3e - a_6] - a_2c'[a_1 + a_2(c + c') + a_3(e + e')]$$
$$+ a_2a_7e' + a_3a_{10}c' - a_7a_{10} \quad (6.73)$$

From the equation $R_{9,1} + eR_{8,2} + e'R_{6,2} = 0$ and the expression for a_{18}, one gets :

$$a_{21} = -a_{10}(ec' + e'c) + ee'(a_6 + a_3e - a_{11})$$
$$+ e^2[a_1 + a_2(c + c') + a_3e - a_6 - a_{11}] \quad (6.74)$$

Finally, the equation $R_{9,3} - R_{8,2} - R_{6,2} = 0$ and the expression for a_{18} yield the equality :

$$a_{23} = 2a_9 + a_6(e - e') + a_{11}(3e + e') + a_{10}(c + c')$$
$$+ a_2(e'c - 2ec - ec') - 2a_1e - 2a_3e^2 - a_3ee' \quad (6.75)$$

The equations $r_{1,1} = r'_{1,1} = R_{5,3} = R_{9,2} = 0$, and (6.63)–(6.75) then prove the claim.

Replacing these seventeen parameters by their expressions in terms of the coordinates $\mathcal{C}$ in the remaining equations leads to an ideal J in the ring $\mathbb{C}[\mathcal{C}]$. Therefore, the variety $R(V)$ is locally isomorphic at (d_o, q_o, d'_o) to the subvariety $V(J)$ of $\mathbb{C}^{13}$. This variety $R(V)$ is locally reducible. The ideal J can indeed be rewritten as a product $J = (h) \cdot L$ where $h = a_1 - a_6 - a_{11} + a_2(c + c') + a_3(e + e')$ and J is the ideal $(c - c', e - e', a_7 - a_3c, a_{10} - a_2e, a_6 - a_{11} - a_2c + a_3e)$ of $\mathbb{C}[\mathcal{C}]$ and the two subvarieties $V((h))$ and $V(L)$ are obviously transverse. The variety $R(V)$ is the union of two irreducible components, one of which is smooth of dimension 12 because it is isomorphic to $V((h))$. The other one is smooth of dimension 8 because it is isomorphic to $V(L)$. As the subvariety $B(V)$ of $R(V)$ is irreducible of dimension 12 (according to proposition 4), $B(V)$ is isomorphic to the smooth hypersurface $V((h))$. The variety $B(V)$ is therefore smooth at $\tilde{q}_o$. The equation $h = 0$ enables us to obtain the coordinate a_1 in terms of the other twelve coordinates. We replace a_1 by its expression in the equations (6.48), (6.48)', (6.63)–(6.75). The variety $B(V)$ is then locally isomorphic to the graph of a morphism :

$$\mathbb{C}^{12} \to \mathbb{C}^{18}$$

$(a_2, a_3, a_5, a_6, a_7, a_9, a_{10}, a_{11}, c, e, c', e') \mapsto (a_1, a_{13}, a_{14}, a_{15}, a_{17}, a_{18}, a_{19}, a_{21}, a_{22}, a_{23},$
$$a, b, d, f, a', b', d', f')$$

and $(a_2, a_3, a_5, a_6, a_7, a_9, a_{10}, a_{11}, c, e, c', e')$ constitutes a local chart.

If now the two doublets d and d' contained in the spherical quadruplet $q_\circ$ do not define the same direction anymore, one can always assume the doublet d to be defined by the ideal $(x^2, y + \alpha x, z)$ where $\alpha \in \mathbb{C}^*$ and the doublet d' is defined by the ideal (x^2, y, z). The following figure represents this configuration :

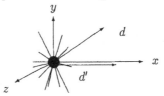

A doublet close to d is then given by the ideal $(x^2 + ax + b, -y + (\mathbf{C} - \alpha)x + d, -z + ex + f)$. In this case, $(a, b, \mathbf{C}, d, e, f)$ constitutes a local chart of $H^2(V)$ at d. One replaces c by $(\mathbf{C} - \alpha)$ everywhere in the previous computations. The variety $R(V)$ is then locally irreducible, smooth of dimension 12 because the subvariety $V(L)$ of $\mathbb{C}^{13}$ is in this case reduced to the empty set – The ideal L of $\mathbb{C}\{a_1, a_2, a_3, a_5, a_6, a_7, a_9, a_{10}, a_{11}, \mathbf{C}, e, c', e'\}$ contains the invertible element $(\mathbf{C} - \alpha - c')$. The inclusion $B(V) \subset R(V)$ is in fact an equality, by proposition 4. This proves the non-singularity of $B(V)$ at each element $(d, q_\circ, d')$. One has proved the following proposition :

Proposition 8 *The variety $B(V)$ is non-singular at every element $(d, q, d') \in B(V)$ where $q \in Hilb^4(V)$ is a spherical quadruplet.*

The proof of theorem 6 is just a consequence of propositions 3, 4, 5 to 8. Let us just note that in the case where either $Res(d, q)$, or $Res(d', q)$ is well defined, it would have been enough to simply compute the dimension of the tangent space of $I(V)$ at (d, q) or at (d', q), in order to prove the non-singularity of $B(V)$. But we were also interested in getting local equations which would be useful for the construction of this variety $\widehat{H^4(V)}$ of quadruplets of V. Beforehand we are going to give a description of the elements of the excess component of the variety $R(V)$.

Description of the elements of the excess component of the variety $R(V)$

The subvariety $R(V)$ of $H^2(V) \times H^4(V) \times H^2(V)$ is the union of two irreducible components, only one of which dominates $H^4(V)$ by the projection Π. This dominant irreducible component is the variety $B(V)$. Let $M(V)$ be the excess component of $R(V)$. The elements of this variety $M(V)$ are elements $\tilde{q} = (d, q, d')$ in $R(V)$ such that :

    - The quadruplet q is a non complete intersection,

    - The doublets d and d' are identical and have the same support that the quadruplet q.

With our drawing conventions, one represents the elements of the variety $M(V)$ as :

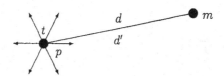

with the extreme configurations :

q elongated q spherical

The variety $M(V)$ is irreducible, smooth of dimension 8. When the support of the quadruplet q is only one point, the variety $M(V)$ is locally isomorphic at $\tilde{q}$ to the subvariety $V(L)$ of $\mathbb{C}^{13}$ (cf. § 6.4.2, p. 95 and 100).

To summarize, we proved in chapter 6 that the closure of the graph of the residual rational map Res defined by :

$$Res : \quad I(V) \quad \cdots \to \quad H^2(V)$$
$$(d, q) \quad \cdots \to \quad d' = q \setminus d$$

is an irreducible, non-singular variety of dimension $4 \cdot \dim V$. In the next chapter, we will use this auxiliary variety to construct the variety $\widehat{H^4(V)}$ as a subvariety of the smooth and irreducible product $[\widehat{H^3(V)}]^4 \times [B(V)]^6$. Recall that the variety $\widehat{H^3(V)}$ denotes the variety of complete triplets of V ([LB1]).

Chapter 7

Construction of the variety $\widehat{H^4(V)}$

Let us now give an explicit construction of the variety $\widehat{H^4(V)}$ of ordered quadruples of V, as a subvariety of the product $[\widehat{H^3(V)}]^4 \times [B(V)]^6$ where $\widehat{H^3(V)}$ denotes the variety of complete triples of V ([LB1]). Let us briefly recall (cf. introduction 0.8) how the element $\hat{q}$ of $\widehat{H^4(V)}$ can be constructed from a generic quadruplet q. Let p^1, p^2, p^3 and p^4 be the four simple points which constitute this quadruplet.

-If t_i denotes the triplet contained in the quadruplet q, disjoint from the simple point p^i, the complete triple in $\widehat{H^3(V)}$ corresponding to the point $(p^1, \ldots, \check{p}^i, \ldots, p^4)$ in V^3 is denoted by $\hat{t}_i$.

-The doublet, which is the union of the two simple points p^i and p^j, is denoted by d_{ij}, the *residual* doublet of d_{ij} in q by d'_{ij}, and the element (d_{ij}, q, d'_{ij}) of $B(V)$ by $\tilde{q}_{ij}$. The element $\hat{q} \in \widehat{H^4(V)}$ constructed from the point (p^1, p^2, p^3, p^4) of V^4 consists of the data $(\hat{t}_1, \hat{t}_2, \hat{t}_3, \hat{t}_4, (\tilde{q}_{ij})_{1 \le i < j \le 4})$ in $[\widehat{H^3(V)}]^4 \times [B(V)]^6$.

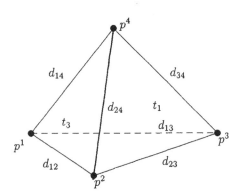

Thus, an element $\hat{q}$ of $\widehat{H^4(V)}$ is an element $(\hat{t}_1, \hat{t}_2, \hat{t}_3, \hat{t}_4, \tilde{q}_{12}, \tilde{q}_{13}, \tilde{q}_{14}, \tilde{q}_{23}, \tilde{q}_{24}, \tilde{q}_{34})$ in the product $[\widehat{H^3(V)}]^4 \times [B(V)]^6$ where

$$\begin{aligned}
\hat{t}_1 &= (\mathbf{p}_2, \mathbf{p}_3, \mathbf{p}_4, \mathbf{d}_{23}, \mathbf{d}_{34}, \mathbf{d}_{24}, t_1) \\
\hat{t}_2 &= (P_1, P_3, P_4, D_{13}, D_{34}, D_{14}, t_2) \\
\hat{t}_3 &= (\mathcal{P}_1, \mathcal{P}_2, \mathcal{P}_4, \mathcal{D}_{12}, \mathcal{D}_{24}, \mathcal{D}_{14}, t_3) \\
\hat{t}_4 &= (p_1, p_2, p_3, d_{12}, d_{23}, d_{13}, t_4) \\
\tilde{q}_{ij} &= (\delta_{ij}, q_{ij}, \delta'_{ij}) \quad \text{for } 1 \le i < j \le 4
\end{aligned}$$

satisfy the conditions :

1. The quadruplets q_{ij} are all equal to a same quadruplet q,

2. One has the equalities : $\delta'_{ij} = \delta_{kl}$ for $\{i, j, k, l\} = \{1, 2, 3, 4\}$

3. The simple points satisfy the equalities :

$$\begin{aligned}
p_1 &= P_1 = \mathcal{P}_1 & (3.i) \\
p_2 &= \mathcal{P}_2 = \mathbf{p}_2 & (3.ii) \\
p_3 &= P_3 = \mathbf{p}_3 & (3.iii) \\
\mathcal{P}_4 &= P_4 = \mathbf{p}_4 & (3.iv)
\end{aligned}$$

4. The doublets satisfy the equalities :

$$\begin{aligned}
d_{12} &= \mathcal{D}_{12} = \delta_{12} \\
d_{23} &= \mathbf{d}_{23} = \delta_{23} \\
d_{13} &= D_{13} = \delta_{13} \\
\mathcal{D}_{24} &= \mathbf{d}_{24} = \delta_{24} \\
\mathcal{D}_{14} &= D_{14} = \delta_{14} \\
D_{34} &= \mathbf{d}_{34} = \delta_{34}
\end{aligned}$$

5. The four triplets t_i are subschemes of the quadruplet q. The points satisfy the conditions : $p_j = Res(t_j, q)$ for $j = 1, 2, 3$ and $\mathbf{p}_4 = Res(t_4, q)$.

Remark 11 : Let $\{p^1, p^2, p^3, p^4\}$ denote the support of a generic quadruplet q of $H^4(V)$. Condition 3 can be rewritten as :

$$\begin{aligned}
p_1 &= P_1 = \mathcal{P}_1 = p^1 & (3.1) \\
p_2 &= \mathcal{P}_2 = \mathbf{p}_2 = p^2 & (3.2) \\
p_3 &= P_3 = \mathbf{p}_3 = p^3 & (3.3) \\
\mathcal{P}_4 &= P_4 = \mathbf{p}_4 = p^4 & (3.4)
\end{aligned}$$

Definition 11 : Let $\widehat{H^4_{\neq}(V)}$ be the open subset of $\widehat{H^4(V)}$, containing the elements $\hat{q}$ having a four-point support.

The study of the variety $\widehat{H^4(V)}$ can be restricted to the case where the support of the quadruplet reduces to one point, since (see property 2, page 69) this variety is already known when the support of the quadruplet contains more than one point : when the support of the quadruplet q consists of two points, the variety $\widehat{H^4(V)}$ is locally isomorphic to one of the products $V \times \widehat{H^3(V)}$ or $'H^2(V) \times 'H^2(V)$:

-If q is the union of a triplet t located at the point p and a simple point m, the data of an element $\hat{q} \in \widehat{H^4(V)}$ is equivalent to the data of an element $(m, \hat{t})$ in the product $V \times \widehat{H^3(V)}$.

-If q is the union of two doublets d_1 and d_2 of respective support p_1 and p_2, the data of an element $\hat{q} \in \widehat{H^4(V)}$ is in this case equivalent to the data of an element $(\tilde{d}_1, \tilde{d}_2)$ in the product $'H^2(V) \times 'H^2(V)$.

In both cases, one deals with a locally smooth (of dimension 12) variety, since it is locally the product of two smooth varieties. Moreover, such elements $\hat{q} \in \widehat{H^4(V)}$ are limits of elements of $\widehat{H^4_{\neq}(V)}$.

In section 7.1, it will be shown that the variety $\widehat{H^4(V)}$ is smooth at each element $\hat{q}$ where q is a locally complete intersection quadruple point and that in the neighborhood of such elements, the open subset $\widehat{H^4_{\neq}(V)}$ is dense in $\widehat{H^4(V)}$. The next section will be devoted to the study of the neighborhood of the elements $\hat{q}$ where q is a non locally complete intersection quadruplet.

7.1 Non-singularity of $\widehat{H^4(V)}$ at $\hat{q}$ where q is a locally complete intersection quadruplet

7.1.1 Case of the curvilinear quadruplet

Let us see what the elements which can be constructed in $\widehat{H^4(V)}$ from an arbitrary curvilinear quadruple point are :

Let q_0 be a curvilinear quadruplet located at the point p. This quadruplet q_0, subscheme of a non-singular curve $\mathcal{C}$, can only contain one triplet : it is the (curvilinear) triplet t of support p, subscheme of $\mathcal{C}$. This quadruplet q_0 also contains only one doublet. We are going to show that in the neighborhood of such an element $\hat{q}_0$, the variety $\widehat{H^4(V)}$ is non-singular of dimension $4 \cdot \dim V$ and that the open subset $\widehat{H^4_{\neq}(V)}$ is dense in $\widehat{H^4(V)}$. The doublets and the triplets contained in the quadruplet q_0 are represented in the following figure :

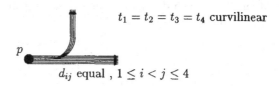

$t_1 = t_2 = t_3 = t_4$ curvilinear

p

d_{ij} equal , $1 \le i < j \le 4$

Let (x, y, z) be a suitable system of local coordinates, centered at p, in which the ideal of q_o is $I(q_o) = (x^4, y, z)$ (see definition 7.(i)). The triplet t contained in this quadruplet is defined by the ideal (x^3, y, z). The ideal (x^2, y, z) defines the doublet d contained in q_o.

Remark 12 Since y and z play exactly the same role, we restrict ourselves to the case where dim $V = 2$. In this case, let (x, y) denote a local coordinate system centered at p.

A quadruplet q_o^{ij} in the neighborhood of q_o is defined by the ideal

$$I(q_o^{ij}) = (x^4 + a^{ij}_{[1]}x^3 + a^{ij}_{[2]}x^2 + a^{ij}_{[3]}x + a^{ij}_{[4]}, \; y + a^{ij}_{[5]}x^3 + a^{ij}_{[6]}x^2 + a^{ij}_{[7]}x + a^{ij}_{[8]})$$

The doublets $\delta_{ij}{}^\circ$ and $\delta_{ij}{}^{\circ'}$ which constitute the element $\widehat{q_o{}^{ij}} = (\delta_{ij}{}^\circ, q_o, \delta_{ij}{}^{\circ'})$ of $B(V)$ are defined by the ideal (x^2, y). The doublets δ_{ij} and δ'_{ij}, in the neighborhood of $\delta_{ij}{}^\circ$ and $\delta_{ij}{}^{\circ'}$ are given by :

$$I(\delta_{ij}) = (x^2 + a_{ij}x + b_{ij}, \; -y + c_{ij}x + d_{ij})$$
$$I(\delta'_{ij}) = (x^2 + a'_{ij}x + b'_{ij}, \; -y + c'_{ij}x + d'_{ij})$$

From (6.17) , the coordinates $(\underline{C}_{ij}) = (a^{ij}_{[1]}, a^{ij}_{[2]}, a^{ij}_{[5]}, a^{ij}_{[6]}, a_{ij}, b_{ij}, c_{ij}, d_{ij})$ constitute a local chart of $B(V)$ at $\widehat{q_o{}^{ij}}$. Using again the notation introduced in [LB1], one defines a triplet t_i in the neighborhood of t by the ideal :

$$I(t_i) = (x^3 + \alpha_i x^2 + \beta_i x + \gamma_i, \; y + \lambda_i x^2 + \mu_i y + \nu_i)$$

Let $(\underline{T_i}) = (s_i, t_i, s'_i, s''_i, c_i, \lambda_i)$ denote the chart of $\widehat{H^3(V)}$ at $\hat{t}_i = (p_j{}^\circ, p_k{}^\circ, p_l{}^\circ, d_{jk}{}^\circ, d_{kl}{}^\circ, d_{jl}{}^\circ, t)$ with $\{i, j, k, l\} = \{1, 2, 3, 4\}$ (see [LB1]). The variety $\widehat{H^3(V)}$ is locally isomorphic at $\hat{t}_i$ to the graph of a morphism from $\mathbb{C}^6$ to $\mathbb{C}^{18}$ (see its expression p.130).

Let us express $\widehat{H^4(V)}$ in the product $[\widehat{H^3(V)}]^4 \times [B(V)]^6$ or more precisely in the $\mathbb{C}^{72}$ of coordinates $(\underline{T}_1, \underline{T}_2, \underline{T}_3, \underline{T}_4, \underline{C}_{12}, \underline{C}_{13}, \underline{C}_{14}, \underline{C}_{23}, \underline{C}_{24}, \underline{C}_{34})$. We are going to show that at $\hat{q}_o$ the variety $\widehat{H^4(V)}$ is isomorphic to the graph of a morphism from $\mathbb{C}^8$ to $\mathbb{C}^{64}$ where $\mathbb{C}^8$ has coordinates $(\underline{C}) = (s_4, t_4, s'_4, s''_4, s''_1, c_4, a^{12}_{[5]}, a^{12}_{[6]})$.

• First, let us show that the sixty four coordinates can be expressed in terms of the other eight.

Condition (3-i), which defines a complete quadruplet, gives the equalities of the co-ordinates of the points :

$$\begin{vmatrix} s_4 \\ t_4 \end{vmatrix} = \begin{vmatrix} s_2 \\ t_2 \end{vmatrix} = \begin{vmatrix} s_3 \\ t_3 \end{vmatrix}$$

Condition (3-ii) can be rewritten as :

$$\begin{vmatrix} s_4 + s'_4 \\ t_4 + t'_4 \end{vmatrix} = \begin{vmatrix} s_3 + s'_3 \\ t_3 + t'_3 \end{vmatrix} = \begin{vmatrix} s_1 \\ t_1 \end{vmatrix}$$

And the expression of t'_4 in the chart $(\mathcal{I}_4)$ is $t'_4 = c_4 s'_4$. Similarly, conditions (3-iii) and (3-iv) can be rewritten as :

$$\begin{vmatrix} s_4 + s'_4 + s''_4 \\ t_4 + t'_4 + t''_4 \end{vmatrix} = \begin{vmatrix} s_2 + s'_2 \\ t_2 + t'_2 \end{vmatrix} = \begin{vmatrix} s_1 + s'_1 \\ t_1 + t'_1 \end{vmatrix}$$

$$\begin{vmatrix} s_3 + s'_3 + s''_3 \\ t_3 + t'_3 + t''_3 \end{vmatrix} = \begin{vmatrix} s_2 + s'_2 + s''_2 \\ t_2 + t'_2 + t''_2 \end{vmatrix} = \begin{vmatrix} s_1 + s'_1 + s''_1 \\ t_1 + t'_1 + t''_1 \end{vmatrix}$$

Thus, conditions (3-i)–(3-iv) enable us to express $s_2, s_3, t_2, t_3, s'_3, s_1, t_1, s'_2, s'_1, s''_3$ and s''_2 in terms of coordinates $\mathcal{C}$. So, one has the expressions

$$s'_3 = s'_4 \tag{7.1}$$

$$s'_1 = s''_4 \tag{7.2}$$

$$s'_2 = s'_4 + s''_4 \tag{7.3}$$

$$s''_2 = s''_1 \tag{7.4}$$

$$s''_3 = s''_1 + s''_4 \tag{7.5}$$

$$t_1 = t_4 + c_4 s'_4 \tag{7.6}$$

From remark 11, let $x[i]$ be the abscissa of the point p^i, contained in the support of the quadruplet q ; one has the change of variables

$$\begin{cases} x[1] = s_4 \\ x[2] = s_4 + s'_4 \\ x[3] = s_4 + s'_4 + s''_4 \\ x[4] = s_4 + s'_4 + s''_4 + s''_1 \end{cases}$$

We will also denote by $(\mathcal{C})$ the set of coordinates $(x[1], x[2], x[3], x[4], t_4, c_4, a^{12}_{[5]},$ $a^{12}_{[6]})$, in order to simplify some of the expressions.

Condition 1 enables us to obtain in particular the following equalities for $1 \leq i < j \leq 4$

$$a^{ij}_{[1]} = a^{12}_{[1]}$$

$$a^{ij}_{[2]} = a^{12}_{[2]}$$

$$a^{ij}_{[5]} = a^{12}_{[5]} \tag{7.7}$$

$$a^{ij}_{[6]} = a^{12}_{[6]} \tag{7.8}$$

In particular, let $a_{[5]} = a_{[5]}^{12}$ and $a_{[6]} = a_{[6]}^{12}$. Conditions 4 lead to the equalities :

$$a_{ij} = -x[i] - x[j] \tag{7.9}$$
$$b_{ij} = x[i]x[j] \tag{7.10}$$

For $\{i, j, k, l\} = \{1, 2, 3, 4\}$, the equalities $a'_{ij} = a_{kl}$ and $b'_{ij} = b_{kl}$ are consequences of condition 2. Moreover, in the chart $(\underline{C}_{ij})$, the expression of a'_{ij} is $a'_{ij} = a_{[1]}^{ij} - a_{ij}$ (equation (6.11), page 86). Using the previous equalities, this equation becomes :

$$-x[k] - x[l] = a_{[1]}^{ij} + (x[i] + x[j])$$

This proves that all the $a_{[1]}^{ij}$ are equal to :

$$a_{[1]} = -(x[1] + x[2] + x[3] + x[4]) \tag{7.11}$$

Similarly, the expression of b'_{ij} in the chart $(\underline{C}_{ij})$ (equation (6.12)) leads to the equality :

$$a_{[2]}^{ij} = x[1]x[2] + x[1]x[3] + x[1]x[4] + x[2]x[3] + x[2]x[4] + x[3]x[4] \tag{7.12}$$

Let $a_{[2]} = a_{[2]}^{12}$. Moreover, from condition 5, which can be expressed by the inclusions of ideals

$$I(p^i) \cdot I(t_i) \subset I(q) \subset I(p^i) \cap I(t_i) \quad,$$

one gets in particular the condition :

$$(*) \qquad (x - x[i])(y + \lambda_i x^2 + \mu_i y + \nu_i) \quad \in \quad I(q)$$

One adds to this element the element $a_{[5]}(x^4 + a_{[1]}x^3 + a_{[2]}x^2 + a_{[3]}x + a_{[4]}) + (x[i] - x)(y + a_{[5]}x^3 + a_{[6]}x^2 + a_{[7]}x + a_{[8]})$ of $I(q)$, in order to obtain the new element R_i :

$$R_i = (-x[i]\nu_i + a_{[4]}a_{[5]} + a_{[8]}x[i]) + (\nu_i + \mu_i x[i] - a_{[8]} + a_{[3]}a_{[5]} + a_{[7]}x[i])x$$
$$+(\mu_i - x[i]\lambda_i - a_{[7]} + a_{[2]}a_{[5]} + a_{[6]}x[i])x^2 + (\lambda_i - a_{[6]} + a_{[1]}a_{[5]} + a_{[5]}x[i])x^3$$

Thus condition $(*)$ is equivalent to the condition $R_i \in I(q)$. From result $[\mathcal{R}es_4]$, § 6.4.0, this condition is equivalent to the condition $R_i = 0$, which implies in particular that the coefficient of x^3 is zero. This yields four equations :

$$\lambda_i = a_{[6]} - a_{[5]}(a_{[1]} + x[i])$$

which can be rewritten using the expression of $a_{[1]}$ (given by (7.11)) as :

$$(e_i) \qquad \lambda_i = a_{[6]} + a_{[5]}(x[j] + x[k] + x[l]) \quad \text{for} \quad \{i, j, k, l\} = \{1, 2, 3, 4\}$$

From condition 4, one gets in particular the equalities of the doublets $\delta_{12} = \mathcal{D}_{12} = d_{12}$, $\delta_{23} = \mathbf{d}_{23} = d_{23}$, $\delta_{13} = d_{13}$. These equalities give in particular the equations :

$$
\begin{cases}
c_{12} = c_4 \\
c_3 = c_4 \\
d_{12} = d_4
\end{cases}
\begin{cases}
c_{23} = c_4 + c'_4 + c''_4 \\
c_1 = c_4 + c'_4 + c''_4 \\
d_{23} = d_4 + d'_4 + d''_4
\end{cases}
\begin{cases}
c_{13} = c_4 + c'_4 \\
c_2 = c_4 + c'_4 \\
d_{13} = d_4 + d'_4
\end{cases}
$$

Warning : In these equations, d_{ij} denotes the sixth coordinate of the doublet δ_{ij}. From the expressions in the chart $(\underline{\mathcal{I}_4})$ of the right hand sides of these equalities and from equation (e_4), one gets $c_{12}, c_3, d_{12}, c_{23}, c_1, d_{23}, c_{13}, c_2$ and d_{13} in terms of $(\underline{\mathcal{C}})$. The equalities of the doublets $\delta_{14} = D_{14}$, $\delta_{24} = d_{24}$ and $\delta_{34} = d_{34}$ give in particular the equations :

$$
\begin{cases} c_{14} = c_2 + c_2' \\ d_{14} = d_2 + d_2' \end{cases} \quad \begin{cases} c_{24} = c_1 + c_1' \\ d_{23} = d_1 + d_1' \end{cases} \quad \begin{cases} c_{34} = c_1 + c_1' + c_1'' \\ d_{34} = d_1 + d_1' + d_1'' \end{cases}
$$

(Here too, the symbol d_{ij} denotes the sixth coordinate of the doublet δ_{ij}.)

Similarly, one replaces the right hand sides of these equalities by their expressions in the chart $(\underline{\mathcal{C}})$. Thus, the sixty four coordinates have been obtained as expressions of the other eight coordinates $s_4, t_4, s_4', s_4'', s_1'', c_4, a_{[5]}, a_{[6]}$.

• Then one checks (using **Macaulay** again) that replacing these sixty four expressions in the remaining equations leads to $0 = 0$. In other words, the sixty four conditions which are necessary to define locally $\widehat{H^4(V)}$ in the product $[\widehat{H^3(V)}]^4 \times [B(V)]^6$ are sufficient. These sixty four expressions enable one to exhibit locally the variety $\widehat{H^4(V)}$ as the graph of a morphism from $\mathbb{C}^8$ to $\mathbb{C}^{64}$ where $\mathbb{C}^8$ has coordinates $(\underline{\mathcal{C}})$. Then the variety $\widehat{H^4(V)}$ is non-singular of dimension $4 \cdot \dim V$ in the neighborhood of $\hat{q}_o$ and $(\underline{\mathcal{C}}) = (s_4, t_4, s_4', s_4'', s_1'', c_4, a_{[5]}, a_{[6]})$ constitutes a local chart. Moreover, the expression of this chart proves that in the neighborhood of $\hat{q}_o$, the open subset $\widehat{H^4_{\neq}(V)}$ is dense in $\widehat{H^4(V)}$.

7.1.2 Case of the square quadruplet

In the next section, we prove the non-singularity of this variety at each element $\hat{q}$ where q is a square quadruplet. We also prove that such elements $\hat{q}$ are limits of elements of the open subset $\widehat{H^4_{\neq}(V)}$.

Starting with such a quadruplet q, let us see what the various elements which can be contructed in $\widehat{H^4(V)}$ are. Such a quadruplet q can only contain one triplet, which is amorphous. From lemma 10, only three of these six doublets contained in the quadruplet can be arbitrary. Recall that if d_α is a doublet of direction α contained in q, the complementary doublet of d_α in q is the doublet of direction $-\alpha$. Of course, in order to prove the non-singularity of $\widehat{H^4(V)}$ at such elements, we will restrict the computations to the most degenerate point $\hat{q}$, i.e. when the six doublets are identical.

Let q_o be a square quadruplet located at a point p of V. Consider a convenient local coordinate system (x, y, z) centered at p in which the quadruplet q_o is defined by the ideal (x^2, y^2, z) of $\mathcal{O}_V$ (cf. definition 7.(ii)) and the six doublets are defined by the same ideal (x^2, y, z). The only triplet t contained in q_o is then defined by the ideal (x^2, xy, y^2, z).

Let $(\mathcal{A}_{ij}) = (a_{ij}, b_{ij}, c_{ij}, d_{ij}, e_{ij}, f_{ij})$ be a chart of $H^2(V)$ at $\delta_{ij}{}^\circ$ and $(\mathcal{A}'_{ij}) = (a'_{ij}, b'_{ij}, c'_{ij}, d'_{ij}, e'_{ij}, f'_{ij})$ be a chart at $\delta_{ij}{}^{\circ\prime}$. One denotes by $(\mathcal{C}_{ij}) = (a^{ij}_{[1]}, a^{ij}_{[3]}, a^{ij}_{[9]}, a^{ij}_{[11]},$ $a_{ij}, b_{ij}, c_{ij}, d_{ij}, e_{ij}, f_{ij}, c'_{ij}, d'_{ij})$ a chart of $B(V)$ at $\widetilde{q_o}^{ij} = (\delta_{ij}{}^\circ, q_o, \delta_{ij}{}^{\circ\prime})$ (cf. remark 10).

If t_i° is the amorphous triplet of ideal (x^2, xy, y^2, z), following the notation introduced in [LB1], one defines a triplet t_i in the neighborhood of t_i° by the ideal :

$$(x^2 + u_i x + v_i y + w_i, \quad xy + u'_i x + v'_i y + w'_i,$$
$$y^2 + u''_i x + v''_i y + w''_i, \quad z + \rho_i x + \sigma_i y + \theta_i)$$

where w_i, w'_i and w''_i are explicit functions of $u_i, v_i, u'_i, v'_i, u''_i, v''_i$ ([I1] , [ELB]). So $(u_i, v_i, u'_i, v'_i, u''_i, v''_i, \rho_i, \sigma_i, \theta_i)$ constitutes a chart of $H^3(V)$ at t_i°. Let us then denote by $(\mathcal{I}_i) = (s_i, t_i, r_i, c_i, c'_i, c''_i, v_i, \rho_i, \sigma_i)$ the chart of $\widehat{H^3}(V)$ at $\hat{t}_i^\circ = (p_j^\circ, p_k^\circ, p_l^\circ, d_{jk}^\circ, d_{kl}^\circ, d_{jl}^\circ, t_i^\circ)$ where $\{i, j, k, l\} = \{1, 2, 3, 4\}$, determined by Le Barz in [LB1] (see A.2).

Now we are going to express locally $\widehat{H^4}(V)$ in the $\mathbb{C}^{108}$ of coordinates $(\mathcal{I}_1, \mathcal{I}_2, \mathcal{I}_3, \mathcal{I}_4, \mathcal{C}_{12}, \mathcal{C}_{13}, \mathcal{C}_{14}, \mathcal{C}_{23}, \mathcal{C}_{24}, \mathcal{C}_{34})$. We will show that at $\hat{q}_o$ the variety $\widehat{H^4}(V)$ is locally isomorphic to the graph of a morphism from $\mathbb{C}^{12}$ to $\mathbb{C}^{96}$ where $\mathbb{C}^{12}$ has coordinates $(\mathcal{C}) = (s_4, t_4, r_4, c_{12}, c_{13}, c_{14}, c_{24}, a^{12}_{[1]}, a^{12}_{[3]}, a^{12}_{[9]}, a^{12}_{[11]}, \rho_4)$.

• First let us show that the ninety six coordinates can be expressed as elements of the algebra $\mathbb{C}\{\mathcal{C}\}$. The equality of all these quadruplets (condition 1) gives in particular the equalities of the coordinates :

$$a^{ij}_{[1]} = a^{12}_{[1]}$$
$$a^{ij}_{[3]} = a^{12}_{[3]}$$
$$a^{ij}_{[9]} = a^{12}_{[9]}$$
$$a^{ij}_{[11]} = a^{12}_{[11]}.$$

From now on, the element $a^{12}_{[l]}$ will be denoted by $a_{[l]}$. Condition 2 yields in particular the equalities of the coordinates : $\begin{cases} c'_{ij} = c_{kl} \\ d'_{ij} = d_{kl} \end{cases}$, for $\{i, j, k, l\} = \{1, 2, 3, 4\}$. The equality of all the coordinates $a^{ij}_{[5]}$ (whose expressions are given in remark 10) provides the equations :

$$a_{[1]} c_{12} c_{34} + c_{12} + c_{34} = a_{[1]} c_{13} c_{24} + c_{13} + c_{24}$$
$$a_{[1]} c_{12} c_{34} + c_{12} + c_{34} = a_{[1]} c_{14} c_{23} + c_{14} + c_{23}$$

These two equalities can be rewritten as :

$$c_{34}(1 + a_{[1]} c_{12}) = a_{[1]} c_{13} c_{24} + c_{13} + c_{24} - c_{12} \qquad (7.13)$$
$$c_{23}(1 + a_{[1]} c_{14}) = a_{[1]} c_{13} c_{24} + c_{13} + c_{24} - c_{14} \qquad (7.14)$$

Since the elements $(1 + a_{[1]} c_{12})$ and $(1 + a_{[1]} c_{14})$ are invertible in the algebra $\mathbb{C}\{\mathcal{C}\}$, these two equalities enable us to obtain the coordinates c_{34} and c_{23} as elements of

$\mathbb{C}\{\mathcal{C}\}$. When one expresses the equalities of condition 4 on the third coordinate of the doublets, one gets the equations :

$$
\begin{cases}
c_1 &= c_{23} \\
c_1' &= c_{24} - c_{23} \\
c_1'' &= c_{34} - c_{24}
\end{cases}
\qquad
\begin{cases}
c_2 &= c_{13} \\
c_2' &= c_{14} - c_{13} \\
c_2'' &= c_{34} - c_{14}
\end{cases}
\qquad
\begin{cases}
c_3 &= c_{12} \\
c_3' &= c_{14} - c_{12} \\
c_3'' &= c_{24} - c_{14}
\end{cases}
\qquad
\begin{cases}
c_4 &= c_{12} \\
c_4' &= c_{13} - c_{12} \\
c_4'' &= c_{23} - c_{13}
\end{cases}
$$

From equations (7.13) and (7.14), the right hand sides of these equalities are elements of $\mathbb{C}\{\mathcal{C}\}$. The equality of the points gives in particular :

$$
(\text{Eq})\qquad
\begin{cases}
s_1 &= s_4 + s_4' \\
t_1 &= t_4 + t_4' \\
r_1 &= r_4' + r_4'
\end{cases}
\qquad\qquad
\begin{cases}
s_2 &= s_3 = s_4 \\
t_2 &= t_3 = t_4 \\
r_2 &= r_3 = r_4
\end{cases}
$$

When one writes the scheme-theoretic inclusions $t_i \subset q$, which are equivalent to the inclusions of ideals $I(q) \subset I(t_i)$, one obtains in particular the conditions :

$$x^2 + a_{[1]}xy + a_{[2]}^{12}x + a_{[3]}y + a_{[4]}^{12} \in I(t_i) \tag{7.15}$$

$$z + a_{[9]}xy + a_{[10]}^{12}x + a_{[11]}y + a_{[12]}^{12} \in I(t_i) \quad . \tag{7.16}$$

Then one divides each of these elements by the ideal $I(t_i)$ (the form of the remainder is given by $[\mathcal{R}es_3]$). Condition (7.15) yields in particular the equalities :

$$(eq_i)\qquad\qquad\qquad a_{[3]} - v_i - a_{[1]}v_i' = 0 \quad .$$

(This term corresponds to the coefficient of y in the expression of the remainder.) Condition (7.16) gives in particular the equations :

$$(E_i)\qquad\qquad\qquad a_{[10]}^{12} - \rho_i - a_9 u_i' = 0$$
$$(F_i)\qquad\qquad\qquad a_{[11]} - \sigma_i - a_9 v_i' = 0$$

(These two expressions correspond to the coefficients of x and y of the remainder of the division.) From equation (eq_i) and the expression of v_i' in the chart $(\mathcal{I}_i)$, one obtains the new equations :

$$(eq_4)'\qquad\qquad v_4(1 + a_{[1]}c_{23}) = a_{[3]} + a_{[1]}s_4$$
$$(eq_3)'\qquad\qquad v_3(1 + a_{[1]}c_{24}) = a_{[3]} + a_{[1]}s_4$$
$$(eq_2)'\qquad\qquad v_2(1 + a_{[1]}c_{34}) = a_{[3]} + a_{[1]}s_4$$
$$(eq_1)'\qquad\qquad v_1(1 + a_{[1]}c_{34}) = a_{[3]} + a_{[1]}(s_4 - v_4(c_{23} - c_{13}))$$

Equation $(eq_3)'$ enables one to express the coordinate v_3 in the ring $\mathbb{C}\{\mathcal{C}\}$, since the element $(1 + a_{[1]}c_{24})$ is invertible. From equations (7.13) and (7.14), the elements $(1+a_{[1]}c_{34})$ and $(1+a_{[1]}c_{23})$ of $\mathbb{C}\{\mathcal{C}\}$ are invertible. Equations $(eq_4)'$ and $(eq_2)'$ enable us to express the coordinates v_4 and v_2 in $\mathbb{C}\{\mathcal{C}\}$. Thus, equation $(eq_1)'$ enables us to obtain the coordinate v_1 as an element of $\mathbb{C}\{\mathcal{C}\}$. From equations (F_i) and the expressions of v'_i in the chart $(\mathcal{C})$, one obtains expressions for the coordinates σ_i. Then, the expressions of $c_4, c''_4, v_4, \sigma_4$ in $\mathbb{C}\{\mathcal{C}\}$ and equations (Eq) enable one to obtain s_1, t_1 and r_1 as elements of $\mathbb{C}\{\mathcal{C}\}$. Equation (E_i) $(i = 1, 2, 3)$ is replaced by $(E_i)' = (E_i) - (E_4)$, which can be rewritten as :

$(E_1)'$ $\qquad\qquad \rho_1 = \rho_4 + a_{[9]}[v_1 c_{23} c_{24} - v_4 c_{12} c_{23}]$

$(E_2)'$ $\qquad\qquad \rho_2 = \rho_4 + a_{[9]}[v_2 c_{13} c_{14} - v_4 c_{12} c_{13}]$

$(E_3)'$ $\qquad\qquad \rho_3 = \rho_4 + a_{[9]}[v_3 c_{12} c_{14} - v_4 c_{12} c_{13}]$

These three equations and the expressions of v_i in $\mathbb{C}\{\mathcal{C}\}$ enable us to obtain expressions for ρ_1, ρ_2 and ρ_3 in the ring $\mathbb{C}\{\mathcal{C}\}$. One writes explicitly the equalities $\delta_{12} = d_{12}$, $\delta_{13} = d_{13}$, $\delta_{23} = d_{23}$, $\delta_{14} = \mathcal{D}_{14}$, $\delta_{24} = \mathcal{D}_{24}$ and $\delta_{34} = D_{34}$ on the coordinates of the doublets. Thus one obtains expressions in $\mathbb{C}\{\mathcal{C}\}$ for the coordinates $a_{ij}, b_{ij}, d_{ij}, e_{ij}, f_{ij}$ of the doublets δ_{ij} since the right hand sides of these equalities are elements of $\mathbb{C}\{\mathcal{C}\}$. Thus, the ninety six coordinates have been obtained as elements of the algebra $\mathbb{C}\{\mathcal{C}\}$.

• Then one checks that when these ninety six expressions are plugged into the remaining equations, one does obtain $0 = 0$. Consequently, the variety $\widehat{H^4(V)}$ is locally isomorphic to the graph of a morphism from $\mathbb{C}^{12}$ to $\mathbb{C}^{96}$ defined in this way. Therefore, the variety $\widehat{H^4(V)}$ is smooth at this point and $(\underline{\mathcal{C}}) = (s_4, t_4, r_4, c_{12}, c_{13}, c_{14}, c_{24}, a_{[1]}, a_{[3]}, a_{[9]}, a_{[11]}, \rho_4)$ constitutes a chart.

Moreover, the above calculations show that this element $\hat{q}_o$ is in the closure of the open subset $\widehat{H^4_{\neq}(V)}$. If $x[i]$ denotes the first coordinate of the point p^i contained in the support of an element $\hat{q}$ in the neighborhood of $\hat{q}_o$, one has the expressions :

$$x[1] = s_4$$
$$x[2] = s_4 - (c_{23} - c_{13})(a_{[3]} + a_{[1]} s_4)[1 + a_{[1]} c_{23}]^{-1}$$
$$x[3] = s_4 - (c_{23} - c_{12})(a_{[3]} + a_{[1]} s_4)[1 + a_{[1]} c_{23}]^{-1}$$
$$x[4] = s_4 - (c_{23} - c_{12})(a_{[3]} + a_{[1]} s_4)[1 + a_{[1]} c_{23}]^{-1} - v_1(c_{24} - c_{23})$$

7.2 The variety $\widehat{H^4(V)}$ at $\hat{q}$ where q is a non locally complete intersection quadruplet

7.2.1 Case of the elongated quadruplet

Consider an elongated quadruplet $q_o \in H^4(V)$ located at a point p of V. According to the definition 7.(iii), the quadruplet q_o is defined by the ideal (x^3, xy, y^2, z) of $\mathcal{O}_V$ for a suitable local coordinate system centered at p.

For such a quadruplet, let us see what the different elements which can be constructed in $\widehat{H^4(V)}$ are :

Recall (see lemma 11) that from such a quadruplet q_o, the elements $\hat{q}_o$ of $B(V)$ all are of the form $\tilde{q}_o = (d, q_o, d')$ where either d or d' is defined by the ideal (x^2, y, z).

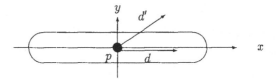

Thus, at an element $\hat{q}_o$ of $\widehat{H^4(V)}$, only three of the six doublets d_{ij}, d_{jl}, d_{li}, where $\{i, j, l\} \subset \{1, 2, 3, 4\}$, can be arbitrary. Then the other three doublets are identical and defined by the ideal (x^2, y, z). The quadruplet q_o can contain the amorphous triplet t of ideal (x^2, xy, y^2, z), as well as the curvilinear triplets t_α defined by the ideals $(x^3, y + \alpha x^2, z)$ where $\alpha \in \mathbb{C}$. The doublets contained in the amorphous triplet t can have an arbitrary direction, while the curvilinear triplet t_α contains only the doublet of ideal (x^2, y, z). We can now give the form of the different elements $\hat{q}_o \in \widehat{H^4(V)}$:

(i) The four triplets t_i are equal and amorphous, only three of the six doublets contained in the quadruplet are arbitrary. The other three doublets are identical and well determined :

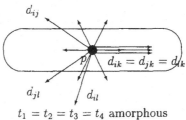

$$t_1 = t_2 = t_3 = t_4 \text{ amorphous}$$

(ii) The quadruplet q_o contains three amorphous triplets t and a curvilinear triplet t_α. The three doublets contained in the triplet t_α are identical and well defined : they all have as ideal (x^2, y, z). The three other doublets contained in the quadruplet q_o can have arbitrary directions :

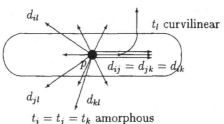

$$t_i = t_j = t_k \text{ amorphous}$$

(iii) The quadruplet q_o contains two amorphous triplets equal to t and two curvilinear triplets. Five of the six doublets (those contained in the curvilinear triplets) are identical and well determined. The sixth doublet contained in q_o has an arbitrary direction :

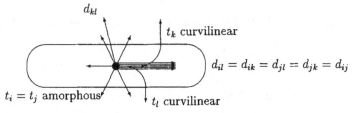

(iv) The quadruplet $q_\circ$ contains an amorphous triplet and three curvilinear triplets. Then the six doublets are identical and well determined :

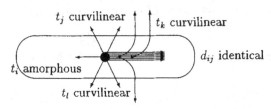

(v) Finally, when the four triplets are curvilinear, the six doublets are then identical and well defined :

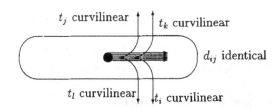

Now, we are going to give local equations of $\widehat{H^4(V)}$ in the product $[\widehat{H^3(V)}]^4 \times [B(V)]^6$ at the most degenerate element $\hat{q}_\circ$, i.e. when the four triplets are amorphous of ideal (x^2, xy, y^2, z), the six doublets are identical and defined by the ideal (x^2, y, z). We will see (proposition 9) that at this point the variety $\widehat{H^4(V)}$ is singular of dimension 12, and that the open subset $\widehat{H^4_{\neq}(V)}$ is dense in $\widehat{H^4(V)}$. In the next section, we will give a geometric description of the singular locus of $\widehat{H^4(V)}$ in the neighborhood of $\hat{q}_\circ$.

A quadruplet q_{ij} in the neighborhood of $q_\circ$ is defined by the ideal :

$$I(q_{ij}) = \quad (x^3 + a_{[1]}^{ij}x^2 + a_{[2]}^{ij}x + a_{[3]}^{ij}y + a_{[4]}^{ij}, \; xy + a_{[5]}^{ij}x^2 + a_{[6]}^{ij}x + a_{[7]}^{ij}y + a_{[8]}^{ij}$$
$$y^2 + a_{[9]}^{ij}x^2 + a_{[10]}^{ij}x + a_{[11]}^{ij}y + a_{[12]}^{ij}, \; z + a_{[13]}^{ij}x^2 + a_{[14]}^{ij}x + a_{[15]}^{ij}y + a_{[16]}^{ij})$$

where $a_{[4]}^{ij}, a_{[8]}^{ij}, a_{[10]}^{ij}$ and $a_{[12]}^{ij}$ are explicit functions of $a_{[1]}^{ij}, a_{[2]}^{ij}, a_{[3]}^{ij}, a_{[5]}^{ij}, a_{[6]}^{ij}, a_{[7]}^{ij}, a_{[9]}^{ij}, a_{[11]}^{ij}$ (see appendix B.1). As usual, let $(\underline{A}_{ij}) = (a_{ij}, b_{ij}, c_{ij}, d_{ij}, e_{ij}, f_{ij})$ – resp. $(\underline{A}'_{ij}) = (a'_{ij}, b'_{ij}, c'_{ij}, d'_{ij}, e'_{ij}, f'_{ij})$ – denote a chart of $H^2(V)$ at $\delta_{ij}{}^\circ$ –resp. at $\delta_{ij}{}^{\circ\prime}$ – where $\delta_{ij}{}^\circ$ and $\delta_{ij}{}^{\circ\prime}$ are defined by the ideal (x^2, y, z). From section 6.4.2, $(\underline{C}_{ij}) = (a_{[1]}^{ij}, a_{[3]}^{ij}, a_{[5]}^{ij}, a_{[6]}^{ij}, a_{[7]}^{ij}, a_{[13]}^{ij}, a_{[14]}^{ij}, a_{[15]}^{ij}, a_{[16]}^{ij}, a_{ij}, c_{ij}, c'_{ij})$ constitutes a chart of $B(V)$ at $\widetilde{q_\circ^{ij}} = (\delta_{ij}{}^\circ, q_\circ, \delta_{ij}{}^{\circ\prime})$. If t_i° is the amorphous triplet of ideal (x^2, xy, y^2, z), and if we use again the notation introduced in [LB1], we define a triplet t_i in the neighborhood of t_i° by the ideal :

$$(x^2 + u_i x + v_i y + w_i, \quad xy + u_i' x + v_i' y + w_i',$$
$$y^2 + u_i'' x + v_i'' y + w_i'', \quad z + \rho_i x + \sigma_i y + \theta_i)$$

where w_i, w_i' and w_i'' are explicit functions of $u_i, v_i, u_i', v_i', u_i'', v_i''$ (see [I1], [ELB]). So $(u_i, v_i, u_i', v_i', u_i'', v_i'', \rho_i, \sigma_i, \theta_i)$ constitutes a chart of $H^3(V)$ at t_i°. Let us then denote by $(\mathcal{I}_i) = (s_i, t_i, r_i, c_i, c_i', c_i'', v_i, \rho_i, \sigma_i)$ the chart of $\widehat{H^3(V)}$ at $\hat{t}_i^\circ = (p_j^\circ, p_k^\circ, p_l^\circ, d_{jk}^\circ, d_{kl}^\circ, d_{jl}^\circ, t_i^\circ)$ where $\{i, j, k, l\} = \{1, 2, 3, 4\}$ (see A.2). Thus we are going to determine the equations of $\widehat{H^4(V)}$ in the $\mathbb{C}^{108}$ of coordinates $(\mathcal{I}_1, \mathcal{I}_2, \mathcal{I}_3, \mathcal{I}_4, \mathcal{C}_{12}, \mathcal{C}_{13}, \mathcal{C}_{14}, \mathcal{C}_{23}, \mathcal{C}_{24}, \mathcal{C}_{34})$.

Remark 13 From remark 11, we denote by $\begin{vmatrix} x[i] \\ y[i] \\ z[i] \end{vmatrix}$ the coordinates of the point p^i.

We perform the change of variables :

$$\begin{cases} x[1] & = & s_4 \\ y[1] & = & t_4 \\ z[1] & = & r_4 \end{cases}$$

We will denote the set of seventeen coordinates by either one of the following notation :

$$\begin{aligned} \mathcal{C} & = (v_1, v_2, v_3, v_4, c_{12}, c_{13}, c_{14}, c_{23}, c_{24}, c_{34}, s_4, t_4, r_4, a_{[5]}^{12}, a_{[13]}^{12}, a_{[14]}^{12}, a_{[15]}^{12}) \\ & = (v_1, v_2, v_3, v_4, c_{12}, c_{13}, c_{14}, c_{23}, c_{24}, c_{34}, x[1], y[1], z[1], a_{[5]}^{12}, a_{[13]}^{12}, a_{[14]}^{12}, a_{[15]}^{12}) \end{aligned}$$

We propose to show the following assertion :

Assertion : the $108 - 17 = 91$ other coordinates are obtained as explicit fonctions of these seventeen coordinates $\mathcal{C}$. Moreover, these seventeen coordinates are linked together by ten linearly independent quadratic equations.

Condition (3.1), which defines a complete quadruplet, gives the equations :

$$\begin{aligned} x[1] & = s_4 = s_3 = s_2 \\ y[1] & = t_4 = t_3 = t_2 \\ z[1] & = r_4 = r_3 = r_2 \end{aligned}$$

Condition 1 in particular gives the equations :

$$a_{[l]}^{ij} = a_{[l]}^{12} \quad \text{for} \quad 1 \leq i < j \leq 4$$
$$\text{and} \quad l \in \{1, 3, 5, 6, 7, 13, 14, 15\}.$$

From now on, we will denote by $a_{[l]}$ the element $a_{[l]}^{12}$ in order to simplify the expressions. Condition 2 ($\delta_{ij}' = \delta_{kl}$) enables us to obtain the expressions :

$$(\star) \qquad c_{ij}' = c_{kl} \quad , \quad \text{for} \quad \{i, j, k, l\} = \{1, 2, 3, 4\}$$

From condition 4, one obtains the equations :

$$(\star\star)\begin{cases} c_1 = c_{23} \\ c'_1 = c_{24} - c_{23} \\ c''_1 = c_{34} - c_{23} \end{cases} \begin{cases} c_2 = c_{13} \\ c'_2 = c_{14} - c_{13} \\ c''_2 = c_{34} - c_{14} \end{cases} \begin{cases} c_3 = c_{12} \\ c'_3 = c_{14} - c_{12} \\ c''_3 = c_{24} - c_{14} \end{cases} \begin{cases} c_4 = c_{12} \\ c'_4 = c_{13} - c_{12} \\ c''_4 = c_{23} - c_{13} \end{cases}$$

In the chart $(\mathcal{I}_4)$ of $\widehat{H^3(V)}$ at $t_4^{\hat{2}}$, the coordinate $x[2]$ is given by the expression :

$$x[2] = x[1] - v_4(c_{23} - c_{13}) \tag{7.17}$$

In the chart $(\mathcal{I}_3)$, one has the expression for $x[2]$:

$$x[2] = x[1] - v_3(c_{24} - c_{14}) \tag{7.18}$$

Similarly, in the different charts $(\mathcal{I}_4), (\mathcal{I}_2)$ and $(\mathcal{I}_1)$ taken in this order, the coordinate $x[3]$ is given by the expression :

$$x[3] = x[1] - v_4(c_{23} - c_{12}) \tag{7.19}$$
$$x[3] = x[1] - v_2(c_{34} - c_{14}) \tag{7.20}$$
$$x[3] = x[2] - v_1(c_{34} - c_{24}) \tag{7.21}$$

Finally, in the different charts $(\mathcal{I}_3), (\mathcal{I}_2)$ and $(\mathcal{I}_1)$ taken in this order, the coordinate $x[4]$ has for expression :

$$x[4] = x[1] - v_3(c_{24} - c_{12}) \tag{7.22}$$
$$x[4] = x[1] - v_2(c_{34} - c_{13}) \tag{7.23}$$
$$x[4] = x[2] - v_1(c_{34} - c_{23}) \tag{7.24}$$

Then one keeps the generators (7.17), (7.19) and (7.22). One replaces the generator (7.18) by the generator $(7.18)' = (7.18) - (7.17)$:

$$v_4(c_{23} - c_{13}) = v_3(c_{24} - c_{14})$$

One replaces the generator (7.20) by $(7.20)' = (7.20) - (7.19)$:

$$v_4(c_{23} - c_{12}) = v_2(c_{34} - c_{14})$$

The generator (7.21) is replaced by $(7.21)' = (7.21) - (7.19) + (7.17)$:

$$v_4(c_{13} - c_{12}) = v_1(c_{34} - c_{24})$$

The generator (7.23) is replaced by $(7.23)' = (7.23) - (7.22)$:

$$v_3(c_{24} - c_{12}) = v_2(c_{34} - c_{13})$$

Finally, one replaces the generator (7.24) by $(7.24)' = (7.24) - (7.22) + (7.18)$:

$$v_3(c_{14} - c_{12}) = v_1(c_{34} - c_{23})$$

The five quadratic equations $(7.18)'$, $(7.20)'$, $(7.21)'$, $(7.23)'$ and $(7.24)'$ are necessary conditions to express the equalities between the points (condition 3).

The scheme-theoretic inclusions $\delta_{ij} \subset t_k$ for $\{i, j, k\} \subset \{1, 2, 3, 4\}$ enable us to obtain in particular the expressions for the coordinates a_{ij} : $a_{ij} = -x[i] - x[j]$.

From the inclusions $I(t_i) \cdot I(p^i) \subset I(q)$ (condition 5), it follows in particular that the generator $(x^2 + u_i x + v_i y + w_i)(x - x[i])$ belongs to the ideal $I(q)$. One adds to this generator an element of $I(q)$ in order to obtain the new element R_i :

$$R_i = x^2(-a_{[1]} - a_{[5]}v_i - x[i] + u_i) + x(-u_i x[i] - w_i - a_{[2]} - a_{[6]}v_i)$$
$$y(-x[i]v_i - a_{[3]} - a_{[7]}v_i) - (x[i]w_i + a_{[4]} + a_{[8]}v_i)$$

Therefore, the condition $(x^2 + u_i x + v_i y + w_i)(x - x[i]) \in I(q)$ is equivalent to the condition $R_i \in I(q)$. From result $[\mathcal{R}es_6]$ one obtains in particular the four equations :

$$a_{[1]} = -a_{[5]}v_i - x[i] + u_i \tag{e_i}$$

We keep equation (e_1), which one rewrites by using the expression of u_1 :

$$a_{[1]} = -3x[1] + v_1(c_{34} - c_{24} - c_{23} - a_{[5]}) + 2v_4(c_{23} - c_{13})$$

One replaces equation (e_2) by $(e_2) - (e_1)$, which one rewrites by using the expressions of $u_1, u_2, x[2]$ and the quadratic equations $(7.18)'$, $(7.23)'$ and $(7.24)'$:

$$v_1(a_{[5]} + c_{24}) = v_2(a_{[5]} + c_{14}) \tag{7.25}$$

One replaces equation (e_3) by $(e_3) - (e_1)$, which one rewrites by using the expressions of $u_1, u_3, x[3]$ and the quadratic equations $(7.18)'$ and $(7.21)'$:

$$v_1(a_{[5]} + c_{23}) = v_3(a_{[5]} + c_{12}) \tag{7.26}$$

Similarly, one replaces equation (e_4) by $(e_4) - (e_1)$, which one rewrites by using the expressions of $u_1, u_4, x[4]$ and the quadratic equations $(7.18)'$ and $(7.24)'$:

$$v_1(a_{[5]} + c_{24}) = v_4(a_{[5]} + c_{12}) \tag{7.27}$$

On the other hand, we have seen (section 6.4.2) that the coordinate a'_{ij} of the doublet δ'_{ij} is given in the chart $\underline{C}_{ij}$ by the expression :

$$a'_{ij} = a_{[1]} + a_{[7]} - a_{ij}$$

The equalities $\begin{cases} a'_{ij} = a_{kl} \\ a_{ij} = -x[i] - x[j] \end{cases}$ lead to the equality :

$$a_{[7]} = -(a_{[1]} + x[1] + x[2] + x[3] + x[4])$$

When $a_{[1]}, x[2], x[3]$ and $x[4]$ are replaced by their expressions in terms of the coordinates $\mathcal{C}$, this equality becomes :

$$a_{[7]} = v_1(a_{[5]} + c_{23}) + v_3(c_{24} - c_{12}) - x[1]$$

We have seen (equation (6.42)) that the expression in the chart $(\mathcal{C}_{ij})$ of the parameters $a_{[9]}^{ij}$ is :

$$a_{[9]}^{ij} = a_{[5]}(c_{ij} + c'_{ij}) + c_{ij}c'_{ij}$$

Using the conditions $c'_{ij} = c_{kl}$, these equalities can be rewritten as :

$$
\begin{aligned}
a_{[9]}^{12} &= a_{[9]}^{34} = a_{[5]}(c_{12} + c_{34}) + c_{12}c_{34} \\
a_{[9]}^{13} &= a_{[9]}^{24} = a_{[5]}(c_{13} + c_{24}) + c_{13}c_{24} \\
a_{[9]}^{23} &= a_{[9]}^{14} = a_{[5]}(c_{14} + c_{23}) + c_{14}c_{23}
\end{aligned}
$$

Then the equality of the coordinates $a_{[9]}^{ij}$ of the quadruplets (consequence of condition 1) gives both quadratic equations :

$$a_{[5]}(c_{12} + c_{34}) + c_{12}c_{34} = a_{[5]}(c_{13} + c_{24}) + c_{13}c_{24} \qquad (7.28)$$

$$a_{[5]}(c_{12} + c_{34}) + c_{12}c_{34} = a_{[5]}(c_{14} + c_{23}) + c_{14}c_{23} \qquad (7.29)$$

The scheme-theoretic inclusions $t_i \subset q$, which are equivalent to the inclusions of ideals $I(q) \subset I(t_i)$, yield in particular the conditions :

$(*)$ $\qquad z + a_{[13]}x^2 + a_{[14]}x + a_{[15]}y + a_{[16]} \quad \in \quad I(t_i)$

One adds to this generator an element of $I(t_i)$ in order to obtain the new element R_i :

$$R_i = x(a_{[14]} - \rho_i - a_{[13]}u_i) + y(a_{[15]} - \sigma_i - a_{[13]}v_i) + (a_{[16]} - \theta_i - w_i a_{[13]})$$

Therefore the conditions $(*)$ are equivalent to the conditions $R_i \in I(t_i)$. From $[\mathcal{R}es_3]$, these conditions can be translated into the equations :

$$
\begin{aligned}
a_{[14]} - \rho_i - a_{[13]}u_i &= 0 & (e_{1i}) \\
a_{[15]} - \sigma_i - a_{[13]}v_i &= 0 & (e_{2i}) \\
a_{[16]} - \theta_i - w_i a_{[13]} &= 0 & (e_{3i})
\end{aligned}
$$

For $i = 1, \ldots, 4$, equation (e_{2i}) enables us to express σ_i in terms of the coordinates $\mathcal{C}$. Similarly, equation (e_{1i}) and the expression of u_i give ρ_i in terms of the coordinates $\mathcal{C}$. The scheme-theoretic inclusion $t_4 \subset q$ gives in particular the two conditions :

$$x^3 + a_{[1]}x^2 + a_{[2]}x + a_{[3]}y + a_{[4]} \quad \in \quad I(t_4) \quad ,$$
$$xy + a_{[5]}x^2 + a_{[6]}x + a_{[7]}y + a_{[8]} \quad \in \quad I(t_4) \quad .$$

The first condition yields in particular the equality :

$$a_{[3]} = v_4(-v'_4 + a_{[1]} - u_4) \quad ,$$

while the second condition gives $\quad a_{[6]} = u'_4 + a_{[5]}u_4$. The expressions which were already obtained for u_4, u'_4, v'_4 and $a_{[1]}$ enable us to express $a_{[3]}$ and $a_{[6]}$ in terms of the coordinates $\mathcal{C}$. Finally, the condition $p^1 \subset q$ gives in particular the equation :

$$z[1] + a_{[13]}x[1]^2 + a_{[14]}x[1] + a_{[15]}y[1] + a_{[16]} = 0$$

which enables us to obtain $a_{[16]}$ in terms of the coordinates $\mathcal{C}$.

Therefore, the ninety one parameters have been obtained as explicit functions of the seventeen coordinates $\mathcal{C}$. Then one verifies that the necessary conditions imposed by the ten quadratic equations $(7.18)'$, $(7.20)'$, $(7.21)'$, $(7.23)'$, $(7.24)'$, (7.25)–(7.29), in order to define the variety $\widehat{H^4(V)}$ locally at $\hat{q}_o$ are sufficient. It suffices to replace all the expressions obtained above in terms of the coordinates $\mathcal{C}$ in the remaining equations and to verify that one obtains either $0 = 0$, either combinations of these ten quadratic equations. One checks that these ten quadratic equations are linearly independent.

Consequently, the variety $\widehat{H^4(V)}$ is locally isomorphic at $\hat{q}_o$ to the subvariety of $\mathbb{C}^{17}$, of coordinates $\mathcal{C}$, defined by these ten quadratic equations. Moreover, the computations show that this element $\hat{q}_o \in \widehat{H^4(V)}$ is the limit of elements of $\widehat{H^4_{\neq}(V)}$. Indeed, if $\hat{q}$ is a complete quadruplet in the neighborhood of $\hat{q}_o$, one obtains the expressions of the first coordinate of the different points of the support of the quadruplet :

$$\begin{aligned}
x[2] &= x[1] - v_4(c_{23} - c_{13}) \\
x[3] &= x[1] - v_4(c_{23} - c_{12}) \\
x[4] &= x[1] - v_3(c_{24} - c_{12})
\end{aligned}$$

and one checks that the differences $x[2] - x[3]$, $x[2] - x[4]$ and $x[3] - x[4]$ are not combinations of the ten previous equations. Next, we prove the proposition :

Proposition 9 *The variety $\widehat{H^4(V)}$ is singular of dimension 12 at $\hat{q}_o$.*

Proof :

We saw that the ten quadratic equations $(7.18)'$, $(7.20)'$, $(7.21)'$, $(7.23)'$, $(7.24)'$, (7.25)–(7.29) constitute local equations of $\widehat{H^4(V)}$ at $\hat{q}_o$ in the ring $\mathbb{C}[\mathcal{C}]$. We perform the change of coordinates :

$$\begin{cases}
X_i &= v_i \qquad\qquad \text{, for } i = 1, \ldots, 4 \\
X_5 &= a_{[5]} + c_{12} \\
X_6 &= a_{[5]} + c_{13} \\
X_7 &= a_{[5]} + c_{14} \\
X_8 &= a_{[5]} + c_{23} \\
X_9 &= a_{[5]} + c_{24} \\
X_{10} &= a_{[5]} + c_{34}
\end{cases}$$

The ideal I generated by the ten quadratic equations can be rewritten in the ring $\mathbb{C}[X_1, \ldots, X_{10}, a_{[5]}, a_{[13]}, a_{[14]}, a_{[15]}, x[1], y[1], z[1]]$ as :

$$
\begin{array}{llll}
X_4 X_5 & = & X_2 X_7 ; & \quad X_4 X_8 = X_2 X_{10} ; \\
X_4 X_5 & = & X_1 X_9 ; & \quad X_2 X_6 = X_3 X_5 ; \\
X_4 X_6 & = & X_3 X_7 ; & \quad X_1 X_8 = X_2 X_6 ; \\
X_4 X_6 & = & X_1 X_{10} ; & \quad X_5 X_{10} = X_6 X_9 ; \\
X_4 X_8 & = & X_3 X_9 ; & \quad X_5 X_{10} = X_7 X_8 ;
\end{array}
$$

These generators are independent of the seven coordinates $a_{[5]}, a_{[13]}, a_{[14]}, a_{[15]}, x[1], y[1]$ and $z[1]$. Therefore the subvariety $V(I)$ of $\mathbb{C}^{17}$ is a product $\widetilde{H} \times \mathbb{C}^7$ where $\widetilde{H}$ is the subvariety of $\mathbb{C}^{10}$ of affine coordinate ring $\mathbb{C}[X_1, \ldots, X_{10}]$, defined by the ideal I, and the coordinates of $\mathbb{C}^7$ are $a_{[5]}, a_{[13]}, a_{[14]}, a_{[15]}, x[1], y[1]$ and $z[1]$. Therefore the variety $\widehat{H^4(V)}$ is locally isomorphic to the product $\widetilde{H} \times \mathbb{C}^7$. In order to prove the proposition, one has to prove that the subvariety $\widetilde{H} \subset \mathbb{C}^{10}$ is singular, of dimension $12 - 7 = 5$. As the ideal I is homogeneous, one must just study the projective subvariety $\mathbb{P}\widetilde{H} \subset \mathbb{P}\mathbb{C}^{10}$ associated to the subvariety $\widetilde{H} \subset \mathbb{C}^{10}$. From now on, the projective space associated to the vector space $\mathbb{C}^{10}$ is denoted by $\mathbb{P}^9$. The affine open subset of $\mathbb{P}^9$ consisting of points of homogeneous coordinates $[X_1 : \cdots : X_{10}]$ such that $X_i \neq 0$ is denoted by U_i. This affine open U_i, of coordinate ring $\mathbb{C}[X_1, \ldots, X_{10}, X_i^{-1}]_0$ is isomorphic to the variety $\mathbb{C}^9$, of coordinate ring $\mathbb{C}[x_1, \ldots, x_{i-1}, x_{i+1}, \ldots, x_{10}]$. (The isomorphism is given by $x_j = \dfrac{X_j}{X_i}$, for $j \neq i$.) We now show that the variety $\mathbb{P}\widetilde{H}$ is singular, of dimension 4, and irreducible. Through the isomorphism $U_i \xrightarrow{\sim} \mathbb{C}^9$, the ideal $I \cap \mathbb{C}[X_1, \ldots, X_{10}, X_i^{-1}]_0$, which defines $\mathbb{P}\widetilde{H} \cap U_i$, is isomorphic to the ideal I_i.

- *Study of $\mathbb{P}\widetilde{H}$ on the open set U_1 :*

The variety $\mathbb{P}\widetilde{H} \cap U_1$ is isomorphic to the subvariety $V(I_1)$ of $\mathbb{C}^9$, where I_1 is the ideal of $\mathbb{C}[x_2, x_3, \ldots, x_{10}]$ generated by the ten following generators :

$$
\begin{align}
x_4 x_5 &= x_2 x_7 \tag{7.30} \\
x_4 x_5 &= x_9 \tag{7.31} \\
x_4 x_6 &= x_3 x_7 \tag{7.32} \\
x_4 x_6 &= x_{10} \tag{7.33} \\
x_4 x_8 &= x_3 x_9 \tag{7.34} \\
x_4 x_8 &= x_2 x_{10} \tag{7.35} \\
x_2 x_6 &= x_3 x_5 \tag{7.36} \\
x_8 &= x_2 x_6 \tag{7.37} \\
x_5 x_{10} &= x_6 x_9 \tag{7.38} \\
x_5 x_{10} &= x_7 x_8 \tag{7.39}
\end{align}
$$

From equations (7.31), (7.33), (7.37), the coordinates x_9, x_{10} and x_8 are obtained as explicit functions of the other coordinates. When one substitutes x_8, x_9 and x_{10} by these expressions into the seven other generators of the ideal I_1, the generator (7.34) becomes proportional to (7.36), the generator (7.39) becomes proportionnal to (7.30) and the equations (7.35) and (7.38) are satisfied. The subvariety $V(I_1)$ of $\mathbb{C}^9$ is isomorphic to the subvariety of $\mathbb{C}^6$, of coordinate ring $\mathbb{C}[x_2, x_3, x_4, x_5, x_6, x_7]$, generated by the three generators (7.30), (7.32) and (7.36). It is nothing else than the determinantal variety defined by :

$$\left\{ M = \begin{pmatrix} x_2 & x_3 & x_4 \\ x_5 & x_6 & x_7 \end{pmatrix} \in M_{2\times 3}(\mathbb{C}) \text{ s.t. } rank M \leq 1 \right\}$$

which is irreducible of dimension 4, as it is shown in the following lemma :

Lemma 13 *The determinantal variety M_1 defined by*

$$M_1 = \left\{ M = \begin{pmatrix} Z_1 & Z_2 & Z_3 \\ Z_4 & Z_5 & Z_6 \end{pmatrix} \in M_{2\times 3}(\mathbb{C}) \text{ s.t. } rank M \leq 1 \right\}$$

is irreducible, of dimension 4, and singular at the origin.

Proof :

The condition $rank M \leq 1$ can be rewritten as :

$$\begin{cases} Z_1 Z_5 &= Z_2 Z_4 \\ Z_1 Z_6 &= Z_3 Z_4 \\ Z_2 Z_6 &= Z_3 Z_5 \end{cases}$$

One easily checks that the projective subvariety $\mathbb{P} M_1 \subset \mathbb{P}^5$ associated to the variety M_1 is irreducible, smooth, of dimension 3. □

On the open set U_1, the variety $\mathbb{P}\widetilde{H}$ is isomorphic to the variety M_1 and is therefore irreducible of dimension 4 and singular at the origin. The origin of this affine open subset U_1 corresponds to the point $A_1 \in \mathbb{P}^9$, of homogeneous coordinates $[1 : 0 : \cdots : 0]$.
• Similarly, one checks that on the nine other open sets U_i of the covering, the variety $\mathbb{P}\widetilde{H}$ is isomorphic to the determinantal variety M_1 and that the singular point of M_1 corresponds to the point A_i of $\mathbb{P}^9$ of homogenous coordinates $[0 : \cdots : 0 : 1 : 0 : \cdots : 0]$ (the 1 is the ith component).

Therefore, one has proved that the variety $\mathbb{P}\widetilde{H}$ is irreducible, of dimension 4, and that its singular locus is the union of the ten points A_i of $\mathbb{P}^9$. Consequently the subvariety $\widetilde{H} \times \mathbb{C}^7 \subset \mathbb{C}^{10} \times \mathbb{C}^7$ is irreducible of dimension $12 = 5+7$, and its singular locus is the union of the ten subvector spaces V_i of $\mathbb{C}^{17}$ where $V_i = V(X_1, \ldots, X_{i-1}, X_{i+1}, \ldots, X_{10})$. As in the neighborhood of $\hat{q}_o$ the variety $\widehat{H^4(V)}$ is locally isomorphic to the product $\widetilde{H} \times \mathbb{C}^7$, it follows that this variety is singular of dimension 12 at $\hat{q}_o$. Therefore, proposition 9 is proved.

In the neighborhood of $\hat{q}_0$, the singular locus $sing(\widehat{H^4(V)})$ of $\widehat{H^4(V)}$ is the union of the ten subvarieties $\mathcal{V}_i \subset \widehat{H^4(V)}$. These ten subvarieties are smooth, of dimension 8. The subvariety $\mathcal{V}_i$ is locally isomorphic to the vector space V_i.

Geometric description of the variety $sing(\widehat{H^4(V)})$ at $\hat{q}_0$:

In this section, we give a geometric description of the singular locus $sing(\widehat{H^4(V)})$ of the variety $\widehat{H^4(V)}$ in the neighborhood of $\hat{q}_0$. Recall that V_i is the subvector space of $\mathbb{C}^{17}$, of dimension 8, defined by the equations: $X_1 = \cdots = X_{i-1} = X_{i+1} = \cdots = X_{10} = 0$. At an element $\hat{q}$ of $sing(\widehat{H^4(V)})$ in the neighborhood of $\hat{q}_0$, the quadruplet q is elongated and its support p has coordinates $(x[1], y[1], z[1])$. Then one performs the change of origin :

$$\begin{cases} \mathcal{X} &= x - x[1] \\ \mathcal{Y} &= x - y[1] \\ \mathcal{Z} &= z - z[1] \end{cases}$$

In the local coordinate system $(\mathcal{X}, \mathcal{Y}, \mathcal{Z})$ centered at p, the quadruplet q is defined by the ideal :

$$I(q) = (\mathcal{X}^3, \mathcal{X}(\mathcal{Y} + a_{[5]}\mathcal{X}), (\mathcal{Y} + a_{[5]}\mathcal{X})^2, \mathcal{Z} + a_{[13]}\mathcal{X}^2 + (a_{[14]} + 2a_{[13]}x[1])\mathcal{X} + a_{[15]}\mathcal{Y})$$

This ideal does define an elongated quadruplet (cf. definition 7.(iii)).

• Geometric description of an element $\hat{q}$ of $\mathcal{V}_i$, for $i = 1, \ldots, 4$:
At such an element, all the doublets and all the triplets have the same support, the point p of V. An element of $\mathcal{V}_1 \cup \mathcal{V}_2 \cup \mathcal{V}_3 \cup \mathcal{V}_4$ satisfies the conditions : $X_5 = \cdots = X_{10} = 0$, which can be rewritten as $c_{ij} = -a_{[5]}$. (The coordinate c_{ij} corresponds to the direction defined by the doublet $d_{ij} \subset q$.) These six equations yield the equality of the six doublets. One checks that these six doublets are defined by the ideal $(\mathcal{X}^2, \mathcal{Y} + a_{[5]}\mathcal{X}, \mathcal{Z} + (a_{[14]} + 2a_{[13]}x[1])\mathcal{X} + a_{[15]}\mathcal{Y})$.
For $j = 1, \ldots, 4$, the condition $X_j = 0$, which can be written as $v_j = 0$, is equivalent to the condition : the triplet t_j is amorphous. Thus, for $i = 1, \ldots, 4$, the triplet t_i is generically curvilinear at an element $\hat{q} \in \mathcal{V}_i$ (since $X_i = v_i$ is generically non zero on V_i). This triplet is defined by the ideal :

$$I(t_i) = (\quad \mathcal{X}^2 + v_i(\mathcal{Y} + a_{[5]}\mathcal{X}), \mathcal{X}\mathcal{Y} - a_{[5]}v_i(\mathcal{Y} + a_{[5]}\mathcal{X}),$$
$$\mathcal{Y}^2 + a_{[5]}^2 v_i(\mathcal{Y} + a_{[5]}\mathcal{X}), \mathcal{Z} + a_{[13]}\mathcal{X}^2 + (a_{[14]} + 2a_{[13]}x[1])\mathcal{X} + a_{[15]}\mathcal{Y})$$

which can be rewritten as follows, if v_i is non zero :

$$I(t_i) = (\mathcal{X}^2 + v_i(\mathcal{Y} + a_{[5]}\mathcal{X}), \mathcal{X}^3, \mathcal{Z} + a_{[13]}\mathcal{X}^2 + (a_{[14]} + 2a_{[13]}x[1])\mathcal{X} + a_{[15]}\mathcal{Y})$$

The other three triplets are amorphous and defined by the ideal :

$$(\mathcal{X}^2, \mathcal{X}\mathcal{Y}, \mathcal{Y}^2, \mathcal{Z} + (a_{[14]} + 2a_{[13]}x[1])\mathcal{X} + a_{[15]}\mathcal{Y})$$

Thus, for $i = 1,\ldots,4$, at a generic element $\hat{q}$ of $\mathcal{V}_i$, the six doublets are identical, the triplet t_i is generically curvilinear and the other three triplets are (identical and) amorphous. On can represent a generic element $\hat{q} \in \mathcal{V}_i$ as :

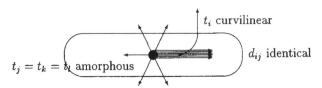

t_i curvilinear

d_{ij} identical

$t_j = t_k = t_l$ amorphous

• Geometric description of an element $\hat{q}$ of $\mathcal{V}_i$, for $i = 5,\ldots,10$:
For $i = 5,\ldots,10$, the elements $\hat{q}$ of $\mathcal{V}_i$ verify the equalities $\quad X_1 = X_2 = X_3 = X_4 = 0$. These equalities, which can be rewritten as $\quad v_1 = v_2 = v_3 = v_4 = 0$, mean that the four triplets are amorphous. These four triplets are defined by the ideal :

$$I(t_i) = (\mathcal{X}^2, \mathcal{X}\mathcal{Y}, \mathcal{Y}^2, \mathcal{Z} + (a_{[14]} + 2a_{[13]}x[1])\mathcal{X} + a_{[15]}\mathcal{Y}) \quad .$$

At an element $\hat{q}$ of $\mathcal{V}_i$, five of the six doublets are identical and defined by the ideal :

$$(\mathcal{X}^2, \mathcal{Y} + a_{[5]}\mathcal{X}, \mathcal{Z} + (a_{[14]} + 2a_{[13]}x[1])\mathcal{X} + a_{[15]}\mathcal{Y})$$

The sixth doublet d_{kl} contained in q (the doublet for which $X_i = a_{[5]} + c_{kl}$ is generically non zero) has an arbitrary direction. This doublet d_{kl} is defined by the ideal :

$$I(d_{kl}) = (\mathcal{X}^2, \mathcal{Y} + c_{kl}\mathcal{X}, \mathcal{Z} + (a_{[14]} + 2a_{[13]}x[1])\mathcal{X} + a_{[15]}\mathcal{Y})$$

Thus, for $i = 5,\ldots,10$, at a generic element $\hat{q}$ of $\mathcal{V}_i$, five of the six doublets are identical, the sixth doublet contained in q has an arbitrary direction and the four triplets are amorphous (and identical). On $\mathcal{V}_5$ (resp. $\mathcal{V}_6, \mathcal{V}_7, \mathcal{V}_8, \mathcal{V}_9$ and $\mathcal{V}_{10}$) it is the doublet d_{12} (resp. $d_{13}, d_{14}, d_{23}, d_{24}$ and d_{34}) which has an arbitrary direction.

Consequently, in the neighborhood of $\hat{q}_o$, the singular locus of $H^4(\widehat{V})$ consists of the following elements :

- The quadruplet q is elongated, the six doublets contained in q are identical, only one of the four triplets contained in the quadruplet is curvilinear. The other three triplets are amorphous :

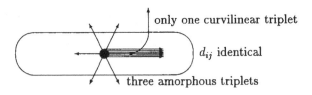

only one curvilinear triplet

d_{ij} identical

three amorphous triplets

- The quadruplet q is elongated, only one of the six doublets contained in q has an arbitrary direction. The other five doublets all define the same direction as that defined by the quadruplet q. The four triplets are amorphous :

only one arbitrary doublet

five of the six doublets identical

the four triplets are amorphous

7.2.2 Case of the spherical quadruplet

Now, let q_o denote the spherical quadruplet of support the point p of V. Such a quadruplet can only contain amorphous triplets. We are going to determine local equations of $\widehat{H^4(V)}$ in the product $[\widehat{H^3(V)}]^4 \times [B(V)]^6$ at the most degenerate element $\hat{q}_o$, i.e. when the four triplets are identical (and amorphous) and the six doublets are identical. In an appropriate system (x, y, z) of local coordinates centered at p, the triplets are defined by the ideal (x^2, xy, y^2, z) and the doublets have for ideal (x^2, y, z). Let $\hat{q}_o = (\hat{t}_1^o, \hat{t}_2^o, \hat{t}_3^o, \hat{t}_4^o, q_o^{12}, q_o^{13}, q_o^{14}, q_o^{23}, q_o^{24}, q_o^{34})$. Let $(\mathcal{T}_i) = (s_i, t_i, r_i, c_i, c'_i, c''_i, v_i, \rho_i, \sigma_i)$ denote again a chart of $\widehat{H^3(V)}$ at $\hat{t}_i^o$ (cf. A.2).

A quadruplet q_o^{ij} in the neighborhood of q_o is defined by the ideal :

$$I(q_o^{ij}) = (\quad x^2 + a_{[1]}^{ij}x + a_{[2]}^{ij}y + a_{[3]}^{ij}z + a_{[4]}^{ij}, \qquad xy + a_{[5]}^{ij}x + a_{[6]}^{ij}y + a_{[7]}^{ij}z + a_{[8]}^{ij},$$
$$xz + a_{[9]}^{ij}x + a_{[10]}^{ij}y + a_{[11]}^{ij}z + a_{[12]}^{ij}, \quad y^2 + a_{[13]}^{ij}x + a_{[14]}^{ij}y + a_{[15]}^{ij}z + a_{[16]}^{ij},$$
$$yz + a_{[17]}^{ij}x + a_{[18]}^{ij}y + a_{[19]}^{ij}z + a_{[20]}^{ij}, \quad z^2 + a_{[21]}^{ij}x + a_{[22]}^{ij}y + a_{[23]}^{ij}z + a_{[24]}^{ij})$$

where $a_{[4]}^{ij}, a_{[8]}^{ij}, a_{[12]}^{ij}, a_{[16]}^{ij}, a_{[20]}^{ij}$ and $a_{[24]}^{ij}$ are explicit functions of the other eighteen parameters, which are denoted by $\underline{a}^{ij}$. These eighteen parameters $\underline{a}^{ij}$ are linked by fifteen quadratic equations, whose expressions are given in appendix B.2. (These fifteen quadratic equations constitute local equations of $H^4(V)$ in the ring $\mathbb{C}[\underline{a}^{ij}]$.)

As previously, $(\mathcal{A}_{ij}) = (a_{ij}, b_{ij}, c_{ij}, d_{ij}, e_{ij}, f_{ij})$ denotes a chart of $H^2(V)$ at δ_{ij}^o, and $(\mathcal{A}'_{ij}) = (a'_{ij}, b'_{ij}, c'_{ij}, d'_{ij}, e'_{ij}, f'_{ij})$ a chart at $\delta_{ij}^{o'}$. Let $(\mathcal{C}_{ij}) = (a_{[2]}^{ij}, a_{[3]}^{ij}, a_{[5]}^{ij}, a_{[6]}^{ij}, a_{[7]}^{ij}, a_{[9]}^{ij}, a_{[10]}^{ij}, a_{[11]}^{ij}, c_{ij}, e_{ij}, c'_{ij}, e'_{ij})$ denote a chart of $B(V)$ at $q_o^{ij} = (\delta_{ij}^o, q_o, \delta_{ij}^{o'})$ (cf. p. 99). We are going to express $\widehat{H^4(V)}$ in the $\mathbb{C}^{108}$, of coordinate ring $\mathbb{C}[\mathcal{T}_1, \mathcal{T}_2, \mathcal{T}_3, \mathcal{T}_4, \mathcal{C}_{ij}]$. Let $(\mathcal{C}) = (s_4, t_4, r_4, \rho_4, \sigma_1, \sigma_2, \sigma_3, \sigma_4, a_{[2]}^{12}, a_{[3]}^{12}, a_{[7]}^{12}, c_{12}, c_{13}, c_{14}, c_{23}, c_{24}, c_{34})$ denote the set of these seventeen coordinates.

Remark 14 From remark 11, if $\begin{vmatrix} x[i] \\ y[i] \\ z[i] \end{vmatrix}$ denotes the coordinates of the point p^i contained in the support of the quadruplet q, one has the change of variables :

$$\begin{cases} x[1] &= s_4 \\ y[1] &= t_4 \\ z[1] &= r_4 \end{cases}$$

The set of the seventeen coordinates :

$$(x[1], y[1], z[1], \rho_4, \sigma_1, \sigma_2, \sigma_3, \sigma_4, a_{[2]}^{12}, a_{[3]}^{12}, a_{[7]}^{12}, c_{12}, c_{13}, c_{14}, c_{23}, c_{24}, c_{34})$$

will be also denoted by $(\mathcal{C})$.

Let us show the following assertion :

Assertion : the $108 - 17 = 91$ other coordinates are explicit functions of the seventeen coordinates $\mathcal{C}$. Moreover, these seventeen coordinates are linked by ten linearly independent equations.

The computations performed in 7.2.1 to express the equalities of the coordinates of the points of V (condition 3) are still valid (equations (7.17)–(7.24)). Therefore, one has again the five quadratic equations (7.18)', (7.20)', (7.21)', (7.23)' and (7.24)'. Condition 2 $(\delta'_{ij} = \delta_{kl})$ enables us to obtain again the equalities $(\star)$ (see page 114), as well as the equalities $e'_{ij} = e_{kl}$, for $\{i, j, k, l\} = \{1, 2, 3, 4\}$. The equations $(\star\star)$ which were established in 7.2.1, page 115 are again satisfied. Condition 1 gives in particular the equalities :

$$a_{[l]}^{ij} = a_{[l]}^{12} \quad \text{for} \quad l \in \{2, 3, 7\} \quad .$$

Then, let $a_{[l]} = a_{[l]}^{12}$, in order to simplify the notation. The scheme-theoretic inclusions $t_i \subset q$ (condition 5), which are equivalent to the inclusion of ideals $I(q) \subset I(t_i)$, yields in particular the condition :

$$x^2 + a_{[1]}^{kl} x + a_{[2]} y + a_{[3]} z + a_{[4]}^{kl} \quad \in \quad I(t_i) \quad .$$

If R_i denotes the remainder of the division of this element by the ideal $I(t_i)$, the previous condition becomes $R_i = 0$. This condition gives in particular the equations :

$$v_i \;=\; a_{[2]} - a_{[3]} \sigma_i \quad ,$$

the element $a_{[2]} - v_i - a_{[3]} \sigma_i$ being the coefficient of y in the remainder R_i.
One substitutes v_1, v_2, v_3 and v_4 by their expressions in terms of the coordinates $\mathcal{C}$ into the equations (7.18)', (7.20)', (7.21)', (7.23)' and (7.24)'. One obtains the new expressions :

$$(a_{[2]} - a_{[3]} \sigma_4)(c_{23} - c_{13}) = (a_{[2]} - a_{[3]} \sigma_3)(c_{24} - c_{14}) \tag{7.40}$$

$$(a_{[2]} - a_{[3]} \sigma_4)(c_{23} - c_{12}) = (a_{[2]} - a_{[3]} \sigma_2)(c_{34} - c_{14}) \tag{7.41}$$

$$(a_{[2]} - a_{[3]} \sigma_4)(c_{13} - c_{12}) = (a_{[2]} - a_{[3]} \sigma_1)(c_{34} - c_{24}) \tag{7.42}$$

$$(a_{[2]} - a_{[3]} \sigma_3)(c_{24} - c_{12}) = (a_{[2]} - a_{[3]} \sigma_2)(c_{34} - c_{13}) \tag{7.43}$$

$$(a_{[2]} - a_{[3]} \sigma_3)(c_{14} - c_{12}) = (a_{[2]} - a_{[3]} \sigma_1)(c_{34} - c_{23}) \tag{7.44}$$

When one expresses condition 4 on the fifth coordinate of the doublets, one obtains the equalities :

$$
\begin{aligned}
e_4 &= & e_3 &= e_{12} \\
e_4 + e_4' + e_4'' &= & e_1 &= e_{23} \\
e_4 + e_4' &= & e_2 &= e_{13} \\
e_3 + e_3' + e_3'' &= & e_1 + e_1' &= e_{24} \\
e_3 + e_3' &= & e_2 + e_2' &= e_{14} \\
e_2 + e_2' + e_2'' &= & e_1 + e_1' + e_1'' &= e_{34}
\end{aligned}
$$

From the six equalities on the left hand side, and from the expressions of e_i, e_i' and e_i'' in the charts $(\mathcal{T}_i)$, one obtains the three expressions :

$$
\begin{aligned}
\rho_1 &= \rho_4 + (\sigma_4 - \sigma_1)c_{23} \\
\rho_2 &= \rho_4 + (\sigma_4 - \sigma_2)c_{13} \\
\rho_3 &= \rho_4 + (\sigma_4 - \sigma_3)c_{12} \quad,
\end{aligned}
$$

as well as the three quadratic equations :

$$
\begin{aligned}
(\sigma_2 - \sigma_3)(c_{14} - c_{13}) &= (\sigma_3 - \sigma_4)(c_{13} - c_{12}) & (7.45) \\
(\sigma_1 - \sigma_2)(c_{34} - c_{13}) &= (\sigma_1 - \sigma_4)(c_{23} - c_{13}) & (7.46) \\
(\sigma_1 - \sigma_3)(c_{24} - c_{23}) &= (\sigma_3 - \sigma_4)(c_{23} - c_{12}) & (7.47)
\end{aligned}
$$

Then, the six equalities on the right hand side enable us to express the coordinates $e_{12}, e_{13}, e_{14}, e_{23}, e_{24}$ and e_{34} in terms of the coordinates $\mathcal{C}$. Moreover, the inclusion of ideals $I(q) \subset I(t_i)$ (condition 5) gives in particular the conditions :

$$
xy + a_{[5]}^{12}x + a_{[6]}^{12}y + a_{[7]}z + a_{[8]}^{12} \in I(t_i).
$$

Let r_i denote the remainder of the division of this element by the ideal $I(t_i)$. The previous conditions become $r_i = 0$. In particular, the coefficients of x and of y of the remainder r_i must be equal to zero, which leads to the equations :

$$
\begin{aligned}
a_{[5]}^{12} &= u_4' + a_{[7]}\rho_4 \\
a_{[6]}^{12} &= v_i' + a_{[7]}\sigma_i
\end{aligned} \qquad (E_i)
$$

The first equation can be rewritten as : $a_{[5]}^{12} = -t_4 - c_{12}c_{13}(a_{[2]} - a_{[3]}\sigma_4) + a_{[7]}\rho_4$. Using the expression of v_4', the equation (E_4) can be rewritten as :

$$
a_{[6]}^{12} = a_{[7]}\sigma_4 + (a_{[2]} - a_{[3]}\sigma_4)c_{23} - s_4
$$

For $j = 1, 2, 3$, one replaces the equation (E_j) by $(E_j)' = (E_j) - (E_4)$, which one rewrites as :

$$
\begin{aligned}
a_{[7]}(\sigma_1 - \sigma_4) + a_{[2]}(c_{34} - c_{13}) + a_{[3]}(\sigma_4 c_{13} - \sigma_1 c_{34}) &= 0 & (7.48) \\
a_{[7]}(\sigma_2 - \sigma_4) + a_{[2]}(c_{34} - c_{23}) + a_{[3]}(\sigma_4 c_{23} - \sigma_2 c_{34}) &= 0 & (7.49) \\
a_{[7]}(\sigma_3 - \sigma_4) + a_{[2]}(c_{24} - c_{23}) + a_{[3]}(\sigma_4 c_{23} - \sigma_3 c_{24}) &= 0 & (7.50)
\end{aligned}
$$

The inclusions of ideals $I(p^i) \cdot I(t_i) \subset I(q)$ give in particular the conditions :

$$(x - x[i])(z + \rho_i x + \sigma_i y + \theta_i) \quad \in \quad I(q).$$

One divides each element by the ideal $I(q)$. The form of the remainder of the division is given by the result $[\mathcal{R}es_7]$. Since this remainder must be equal to zero, one has in particular the four equations :

$$a^{12}_{[11]} = -x[i] - a_{[7]}\sigma_i - a_{[3]}\rho_i \qquad (eq_i)$$

(This expression corresponds to the coefficient of z in the remainder.) Equation (eq_1) and the expression of ρ_1 in the chart $\mathcal{C}$ enable one to obtain an expression for the coordinate $a^{12}_{[11]}$ in terms of the coordinates $\mathcal{C}$. For $j = 2, 3, 4$, one replaces equation (eq_j) by $(eq_j)' = (eq_j) - (eq_1)$, which one rewrites as :

$$(\sigma_2 - \sigma_1)(a_{[7]} - a_{[3]}c_{34}) = (a_{[2]} - a_{[3]}\sigma_4)(c_{23} - c_{13}) \qquad (7.51)$$

$$(\sigma_3 - \sigma_1)(a_{[7]} - a_{[3]}c_{24}) = (a_{[2]} - a_{[3]}\sigma_4)(c_{23} - c_{12}) \qquad (7.52)$$

$$(\sigma_4 - \sigma_1)(a_{[7]} - a_{[3]}c_{23}) = (a_{[2]} - a_{[3]}\sigma_2)(c_{34} - c_{13}) \qquad (7.53)$$

The condition $xz + a^{12}_{[9]}x + a^{12}_{[10]}y + a^{12}_{[11]}z + a^{12}_{[12]} \in I(t_4)$, which is a consequence of the inclusion $I(q) \subset I(t_4)$, enables us to obtain similar expressions :

$$a^{12}_{[9]} = a^{12}_{[11]}\rho_4 - \sigma_4 u'_4 - u_4\rho_4 + \theta_4 \quad ,$$
$$a^{12}_{[10]} = -v_4\rho_4 - \sigma_4 v'_4 + a^{12}_{[11]}\sigma_4 \quad .$$

One replaces the right hand side terms by their expressions in terms of the coordinates $\mathcal{C}$, which enables one to obtain the expressions for $a^{12}_{[9]}$ and $a^{12}_{[10]}$. From the equation (6.56), p. 97 and the equalities $c'_{ij} = c_{kl}$ $(\star)$, the expressions

$$a^{ij}_{[15]} = -a_{[3]}c_{ij}c_{kl} + a_{[7]}(c_{ij} + c_{kl})$$

follow. The equality of all the coordinates $a^{ij}_{[15]}$ (condition 1) gives the two equations :

$$a_{[7]}(c_{12} + c_{34}) - a_{[3]}c_{12}c_{34} = a_{[7]}(c_{13} + c_{24}) - a_{[3]}c_{13}c_{24} \qquad (7.54)$$

$$a_{[7]}(c_{12} + c_{34}) - a_{[3]}c_{12}c_{34} = a_{[7]}(c_{14} + c_{23}) - a_{[3]}c_{14}c_{23} \qquad (7.55)$$

The ninety one coordinates have been obtained explicitly in terms of the other seventeen $\mathcal{C}$. One checks that one obtains either $0 = 0$, or combinations of the sixteen equations $(7.40) - (7.55)$ in the ring $\mathbb{C}[\mathcal{C}]$, when these ninety one expressions are substituted in the remaining equations. Then one verifies that the ideal I of $\mathbb{C}[\mathcal{C}]$, generated by these sixteen equations, can in fact be generated by the ten linearly independent equations :

$$(\sigma_3 - \sigma_4)c_{12} + (-\sigma_2 + \sigma_4)c_{13} + (\sigma_2 - \sigma_3)c_{14} = 0$$
$$(\sigma_2 - \sigma_4)c_{13} + (-\sigma_1 + \sigma_4)c_{23} + (\sigma_1 - \sigma_2)c_{34} = 0$$
$$(\sigma_2 - \sigma_3)c_{14} + (-\sigma_1 + \sigma_3)c_{24} + (\sigma_1 - \sigma_2)c_{34} = 0$$
$$(c_{13} - c_{14} - c_{23} + c_{24})a_{[2]} + (-\sigma_4 c_{13} + \sigma_3 c_{14} + \sigma_4 c_{23} - \sigma_3 c_{24})a_{[3]} = 0$$
$$(c_{12} - c_{14} - c_{23} + c_{34})a_{[2]} + (-\sigma_4 c_{12} + \sigma_4 c_{23} + \sigma_1 c_{24} - \sigma_3 c_{24} - \sigma_1 c_{34})a_{[3]} = 0$$
$$(c_{14} + c_{23} - c_{24} - c_{34})a_{[2]} + (-\sigma_3 c_{14} - \sigma_4 c_{23} + \sigma_3 c_{24} + \sigma_1 c_{34})a_{[3]} + a_{[7]}(-\sigma_1 + \sigma_4) = 0$$
$$(c_{23} - c_{34})a_{[2]} + (-\sigma_4 c_{23} + \sigma_2 c_{34})a_{[3]} + (-\sigma_2 + \sigma_4)a_{[7]} = 0$$
$$(c_{24} - c_{34})a_{[2]} + (-\sigma_3 c_{24} + \sigma_2 c_{34})a_{[3]} + (-\sigma_2 + \sigma_3)a_{[7]} = 0$$
$$(c_{13}c_{24} - c_{12}c_{34})a_{[3]} + (c_{12} - c_{13} - c_{24} + c_{34})a_{[7]} = 0$$
$$(c_{14}c_{23} - c_{12}c_{34})a_{[3]} + (c_{12} - c_{14} - c_{23} + c_{34})a_{[7]} = 0$$

We have shown the assertion.

The variety $\widehat{H^4(V)}$ is locally isomorphic at $\hat{q}_o$ to the subvariety of $\mathbb{C}^{17}$ of coordinates $(\mathcal{C})$, defined by these ten equations. Since the generators of I are independent of the coordinates $x[1], y[1], z[1]$ and ρ_4, consequently the variety $\widehat{H^4(V)}$ is locally isomorphic to the product $V(I) \times \mathbb{C}^4 \subset \mathbb{C}^{13} \times \mathbb{C}^4$, where I is in this case considered as an ideal of the ring $\mathbb{C}[\sigma_1, \sigma_2, \sigma_3, \sigma_4, a_{[2]}, a_{[3]}, a_{[7]}, c_{12}, c_{13}, c_{14}, c_{23}, c_{24}, c_{34}]$. One checks that the subvariety $V(I)$ of $\mathbb{C}^{13}$ is irreducible, singular, of dimension 8. As a consequence, the variety $\widehat{H^4(V)}$ is singular of dimension $12(= 8 + 4)$ at the point $\hat{q}_o$. The local equations of $\widehat{H^4(V)}$ being in this case slightly more complicated, we have not been able so far to obtain a geometric description of the singular locus $sing(\widehat{H^4(V)})$ in the neighborhood of this element $\hat{q}_o$. However, the computations enable us to prove that in the neighborhood of this element $\hat{q}_o$, the open set $H^4_{\neq}(V)$ is dense. The first coordinate $x[i]$ of the point p^i contained in the support of an element close to $\hat{q}_o$ has for expression :

$$x[2] = x[1] - (a_{[2]} - a_{[3]}\sigma_4)(c_{23} - c_{13})$$
$$x[3] = x[1] - (a_{[2]} - a_{[3]}\sigma_4)(c_{23} - c_{12})$$
$$x[4] = x[1] - (a_{[2]} - a_{[3]}\sigma_2)(c_{34} - c_{13})$$

and one checks that the differences $(x[2] - x[3]), (x[2] - x[4]), (x[3] - x[4])$ are not elements of the ideal I.

7.3 Irreducibility of $\widehat{H^4(V)}$

We saw in the introduction of this chapter that when the support of the element $\hat{q} \in \widehat{H^4(V)}$ contains at least two points, the variety $\widehat{H^4(V)}$ is locally isomorphic to one of the two products $'H^2(V) \times' H^2(V), V \times \widehat{H^3(V)}$. Therefore such an element $\hat{q}$ is the limit of elements of the open subset $H^4_{\neq}(V)$. The computations performed in sections 7.1 and 7.2 show that when q is a quadruple point, the element $\hat{q} \in \widehat{H^4(V)}$

is again in the closure of the open set $\widehat{H^4_{\neq}(V)}$. (See sections 7.1 and 7.2 for explicit deformations of a quadruple point into a simple quadruplet.) Consequently, the open set $\widehat{H^4_{\neq}(V)}$ is dense in $\widehat{H^4(V)}$. The variety constructed in this way is irreducible, of dimension $12 (= 4 \cdot \dim V)$.

Furthermore, the projection :

$$\widehat{H^4(V)} \quad \to \quad H^4(V)$$
$$\hat{q} \quad \mapsto \quad q$$

is generically a 4!-sheeted cover, because the open set $H^4_{\neq}(V)$ is dense in $H^4(V)$ ([F], [I2]).

Appendix A

In this appendix, one recalls the local equations of the variety of complete triples of V, derived in [LB1].

A.1 Local chart of $\widehat{H^3(V)}$ at $\hat{t}$, where t is a curvilinear triple point

Assume that the variety V is of dimension 2. Let us denote by t_4° a curvilinear triplet of support the point p of V. Such a triplet can only contain one doublet d. In the neighborhood of the complete triplet $\hat{t_4^\circ} = (p, p, p, d, d, d, t_4^\circ)$, the variety $\widehat{H^3(V)}$ is isomorphic to the graph of a morphism from $\mathbb{C}^6$ to $\mathbb{C}^{18}$. We recall here its expression.

In an appropriate local coordinate system (x, y) centered at p, the triplet t_4° is defined by the ideal (x^3, y). Therefore the doublet d is defined by the ideal (x^2, y). A triplet t_4 close to t_4° is defined by the ideal :

$$I(t_4) = (x^3 + \alpha_4 x^2 + \beta_4 x + \gamma_4, \, y + \lambda_4 x^2 + \mu_4 x + \nu_4) \quad .$$

One defines the doublets d_{12}, d_{23} and d_{13} close to d respectively by the ideals :

$$I(d_{12}) = (x^2 + a_4 x + b_4, \, -y + c_4 x + d_4)$$
$$I(d_{13}) = (x^2 + (a_4 + a_4')x + b_4 + b_4', \, -y + (c_4 + c_4')x + d_4 + d_4')$$
$$I(d_{23}) = (x^2 + (a_4 + a_4' + a_4'')x + b_4 + b_4' + b_4'', \, -y + (c_4 + c_4' + c_4'')x + d_4 + d_4' + d_4'')$$

The points p_1, p_2 and p_3 in the neighborhood of p have coordinates

$$p_1 \begin{vmatrix} s_4 \\ t_4 \end{vmatrix} \quad p_2 \begin{vmatrix} s_4 + s_4' \\ t_4 + t_4' \end{vmatrix} \quad p_3 \begin{vmatrix} s_4 + s_4' + s_4'' \\ t_4 + t_4' + t_4'' \end{vmatrix}$$

In the neighborhood of the complete triplet $\hat{t_4^\circ}$, the variety $\widehat{H^3(V)}$ is isomorphic to the graph of a morphism ([LB1] pp 941-944) :

$$\mathbb{C}^6 \quad \to \quad \mathbb{C}^{18}$$
$$(s_4, t_4, s_4', s_4'', c_4, \lambda_4) \mapsto (t_4', t_4'', a_4, b_4, d_4, a_4', b_4', c_4', d_4', a_4'', b_4'', c_4'', d_4'', \alpha_4, \beta_4, \gamma_4, \mu_4, \nu_4)$$

defined by :

$$
\begin{aligned}
t_4' &= c_4 s_4' & a_4'' &= -s_4' \\
t_4'' &= -s_4''(c_4 - \lambda_4(s_4' + s_4'')) & b_4'' &= s_4'(s_4 + s_4' + s_4'') \\
a_4 &= -2s_4 - s_4' & c_4'' &= -\lambda_4 s_4' \\
b_4 &= s_4(s_4 + s_4') & d_4'' &= \lambda_4 s_4'(s_4 + s_4' + s_4'') \\
d_4 &= t_4 - c_4 s_4 & \alpha_4 &= -(3s_4 + 2s_4' + s_4'') \\
a_4' &= -s_4'' & \beta_4 &= s_4(s_4 + s_4') + (2s_4 + s_4')(s_4 + s_4' + s_4'') \\
b_4' &= s_4 s_4'' & \gamma_4 &= -s_4(s_4 + s_4')(s_4 + s_4' + s_4'') \\
c_4' &= -\lambda_4 s_4'' & \mu_4 &= -c_4 + \lambda_4(2s_4 + s_4') \\
d_4'' &= \lambda_4 s_4 s_4'' & \nu_4 &= \lambda_4 s_4(s_4 + s_4') - t_4 + c_4 s_4
\end{aligned}
$$

Remark 15 : We have assumed that the variety V is of dimension 2. For V of dimension n, it suffices to replace the coordinate system (x, y) of V by $(x, \vec{y})$, where $\vec{y} = (y_1, \ldots, y_{n-1})$ (recall notation 18, p. 44). In this case, the ideal of d is $(x^2, \vec{y})$ and the triplet t_4° has for ideal $(x^3, \vec{y})$. Then a triplet t_4 in the neighborhood of t_4° is given by the ideal :

$$
I(t_4) = (x^3 + \alpha_4 x^2 + \beta_4 x + \gamma_4 , \, \vec{y} + \vec{\lambda}_4 x^2 + \vec{\mu}_4 x + \vec{\nu}_4) \quad .
$$

Similarly, the doublets d_{12}, d_{23} and d_{13} in the neighborhood of d are defined by the ideals

$$
\begin{aligned}
I(d_{12}) &= (x^2 + a_4 x + b_4 , \, -\vec{y} + \vec{c}_4 x + \vec{d}_4) \\
I(d_{13}) &= (x^2 + (a_4 + a_4')x + b_4 + b_4' , \, -\vec{y} + (\vec{c}_4 + \vec{c}_4')x + \vec{d}_4 + \vec{d}_4') \\
I(d_{23}) &= (x^2 + (a_4 + a_4' + a_4'')x + b_4 + b_4' + b_4'' , \, -\vec{y} + (\vec{c}_4 + \vec{c}_4' + \vec{c}_4'')x + \vec{d}_4 + \vec{d}_4' + \vec{d}_4'') \quad .
\end{aligned}
$$

A local chart of $\widehat{H^3(V)}$ at $\hat{t}_4^\circ$ is in this case given by $(s_4, \vec{t}_4, s_4', s_4'', \vec{c}_4, \vec{\lambda}_4)$.

A.2 Local chart of $\widehat{H^3(V)}$ at $\hat{t}$, where t is amorphous

One recalls here the local equations of the variety $\widehat{H^3(V)}$ in the neighborhood of a complete triplet $\hat{t}_4^\circ = (p_1^\circ, p_2^\circ, p_3^\circ, d_{12}^\circ, d_{23}^\circ, d_{13}^\circ, t_4^\circ)$ where t_4° is an amorphous triplet of support the point p of V and the three doublets are identical. In an appropriate local coordinate system (x, y, z) centered at p, the triplet t_4° is defined by the ideal (x^2, xy, y^2, z) and the doublets are defined by the same ideal (x^2, y, z). Then a triplet t_4 close to t_4° is given by the ideal

$$
\begin{aligned}
(x^2 + u_4 x + v_4 y + w_4, \quad &xy + u_4' x + v_4' y + w_4', \\
&y^2 + u_4'' x + v_4'' y + w_4'', \quad z + \rho_4 x + \sigma_4 y + \theta_4) \quad ,
\end{aligned}
$$

where w_4, w_4' and w_4'' verify the relationships :

$$\begin{cases} w_4 &= u_4 v_4' + v_4 v_4'' - u_4' v_4 - v_4'^2 \\ w_4' &= u_4' v_4' - v_4 u_4'' \\ w_4'' &= u_4 u_4'' + u_4' v_4'' - u_4'^2 - u_4'' v_4' \end{cases}$$

Using the notation introduced in [LB1], one defines the doublets d_{ij} close to d_{ij}° by the different ideals :

$$\begin{aligned} I(d_{12}) &= (x^2 + a_4 x + b_4, -y + c_4 x + d_4, -z + e_4 x + f_4) \\ I(d_{13}) &= (x^2 + (a_4 + a_4')x + b_4 + b_4', -y + (c_4 + c_4')x + d_4 + d_4', \\ & \quad -z + (e_4 + e_4')x + f_4 + f_4') \\ I(d_{23}) &= (x^2 + (a_4 + a_4' + a_4'')x + b_4 + b_4' + b_4'', -y + (c_4 + c_4' + c_4'')x + d_4 + d_4' + d_4'', \\ & \quad -z + (e_4 + e_4' + e_4'')x + f_4 + f_4' + f_4'') \end{aligned}$$

The coordinates of the points p_1, p_2, p_3 in the neighborhood of p are :

$$p_1 \begin{vmatrix} s_4 \\ t_4 \\ r_4 \end{vmatrix} \quad p_2 \begin{vmatrix} s_4 + s_4' \\ t_4 + t_4' \\ r_4 + r_4' \end{vmatrix} \quad p_3 \begin{vmatrix} s_4 + s_4' + s_4'' \\ t_4 + t_4' + t_4'' \\ r_4 + r_4' + r_4'' \end{vmatrix}$$

In the neighborhood of a complete triplet $\widehat{t_4^\circ}$, the variety $\widehat{H^3(V)}$ is isomorphic to the graph of a morphism from $\mathbb{C}^9$ to $\mathbb{C}^{27}$ (c.f. [LB1] pp 934-938) :

$$\mathbb{C}^9 \qquad \rightarrow \quad \mathbb{C}^{27}$$
$$(s_4, t_4, r_4, c_4, c_4', c_4'', v_4, \rho_4, \sigma_4) \mapsto (s_4', t_4', r_4', s_4'', t_4'', r_4'', a_4, b_4, d_4, e_4, f_4, a_4', b_4', d_4', e_4', f_4',$$
$$a_4'', b_4'', d_4'', e_4'', f_4'', u_4, u_4', v_4', u_4'', v_4'', \theta_4)$$

Its expression is :

$$\begin{aligned} s_4' &= -c_4'' v_4 \\ t_4' &= -c_4 c_4'' v_4 \\ r_4' &= c_4'' v_4 (\rho_4 + \sigma_4 c_4) \\ s_4'' &= -c_4' v_4 \\ t_4'' &= -c_4' v_4 (c_4 + c_4' + c_4'') \\ r_4'' &= c_4' v_4 (\rho_4 + \sigma_4(c_4 + c_4' + c_4'')) \\ a_4 &= -2s_4 + c_4'' v_4 \\ b_4 &= s_4(s_4 - c_4'' v_4) \\ d_4 &= t_4 - c_4 s_4 \\ e_4 &= -\rho_4 - \sigma_4 c_4 \\ f_4 &= r_4 + s_4(\rho_4 + \sigma_4 c_4) \\ a_4' &= c_4' v_4 \\ b_4' &= -s_4 c_4' v_4 \\ d_4' &= -c_4' s_4 \end{aligned}$$

$$\begin{aligned} e_4' &= -\sigma_4 c_4' \\ f_4' &= s_4 \sigma_4 c_4' \\ a_4'' &= c_4' v_4 \\ b_4'' &= -c_4' v_4 (s_4 - v_4(c_4' + c_4'')) \\ d_4'' &= c_4'(-s_4 + v_4(c_4' + c_4'')) \\ e_4'' &= -\sigma_4 c_4'' \\ f_4'' &= \sigma_4 c_4''(s_4 - v_4(c_4' + c_4'')) \\ u_4 &= -2s_4 + v_4(c_4'' - c_4) \\ u_4' &= -t_4 - v_4 c_4(c_4 + c_4') \\ v_4' &= v_4(c_4 + c_4' + c_4'') - s_4 \\ u_4'' &= -c_4 v_4(c_4 + c_4')(c_4 + c_4' + c_4'') \\ v_4'' &= -2t_4 + v_4(c_4 c_4'' + (c_4 + c_4')(c_4 + c_4' + c_4'')) \\ \theta_4 &= -r_4 - \rho_4 s_4 - \sigma_4 t_4 \end{aligned}$$

Appendix B

According to [BGS], [Gr], [I1], it is always possible to determine explicitly the neighborhood of $H^n(V)$ at a n-uple point ξ, which one identifies with the flattener of a germ of an application. (We have at our disposal the division theorem with parameters of Galligo ([Gal]) to compute explicitly the flattener of a germ of an application.)

In this section, we give local equations of the Hilbert scheme $H^4(V)$ in the neighborhood of non-locally complete intersections quadruplets. Recall that we have assumed dim $V = 3$. Then, let ξ be a point of the Hilbert scheme $H^n(V)$, of support the point p of V. The ideal of $\mathcal{O}_V$ which defines this n-uplet is denoted by $I(\xi)$. One denotes by (x, y, z) a local coordinate system centered at p. Briançon, Granger and Speder (cf. [BGS], [Gr], [I1]) have found a method to obtain explicitly local equations of $H^n(V)$ at ξ :

One denotes by $(f_1, \ldots, f_p)$ the standard basis of the ideal $I(\xi)$ (once the direction has been chosen). The quotient $\dfrac{\mathcal{O}_V}{(f_1, \ldots, f_p)}$ is a $\mathbb{C}$-vector space of dimension n. One denotes by $\Delta = \{\overline{e_1}, \ldots, \overline{e_n}\}$ a basis of this $\mathbb{C}$-vector space where $e_1, \ldots, e_n$ are elements of $\mathcal{O}_V$. Let $F_i = f_i + \sum_{j=1}^n a_{ij} e_j$, for $i = 1, \ldots, n$, where all the parameters a_{ij} are elements of $\mathbb{C}$. The set of these pn parameters is denoted by $(\underline{a})$. Let $\mathcal{O}_S = \mathbb{C}\{\underline{a}\}$. The family $(F_1, \ldots, F_p)$ defines a morphism at the origin of $\mathbb{C}^3 \times \mathbb{C}^{pn}$ parametred by $(x, y, z), (\underline{a})$. In the neighborhood of the origin, one considers the subvariety $\mathcal{X}$ of $\mathbb{C}^3 \times \mathbb{C}^{pn}$ defined by the cancellation of the polynomials $F_1(x, y, z, \underline{a}), \ldots, F_p(x, y, z, \underline{a})$. This subvariety $\mathcal{X} = V((F_1, \ldots, F_p))$ possesses a projection onto $\mathbb{C}^{pn}$, which is the restriction to $\mathcal{X}$ of the projection from the product $\mathbb{C}^3 \times \mathbb{C}^{pn}$ onto the second factor :

$$
\begin{array}{cc}
(\mathcal{X}, 0) & \subset \quad (\mathbb{C}^3 \times \mathbb{C}^{pn}, 0) \\
\downarrow \Pi & \\
(\mathbb{C}^{pn}, 0) &
\end{array}
$$

Briançon, Granger and Speder ([BGS], [Gr], [I1]) have shown that the Hilbert scheme $H^n(V)$ is locally isomorphic at ξ to the flattener of Π. The equations of this flattener can be obtained in the following manner, according to [Gal] – proposition 1.4.7 :

One constructs a basis $\{(E_j) = (E_{1j}, \ldots, E_{pj})\}$ of relations between the f_i. Therefore, for each j, one has the equality :

$$\sum_{i=1}^{p} E_{ij} f_i = 0 \quad .$$

One creates in $\mathcal{O}_S\{x, y, z\}$ the products :

$$M_j = \sum_{i=1}^{p} E_{ij} F_i \quad .$$

Then one constructs the remainders of the division of the M_j by the family $(F_1, \ldots, F_p)$ (cf. [Gal] theorem 1.2.7). One has :

$$H_j = \sum_{i=1}^{n} h_{ji}(\underline{a}) e_i \quad .$$

(This remainder is not zero if $\dfrac{\mathcal{O}_S\{x, y, z\}}{(F_1, \ldots, F_p)}$ is not a flat $\mathcal{O}_S$-module.)
The equations $\{h_{ji}(\underline{a}) = 0\}$ are the equations in $\mathcal{O}_S$ which define $H^n(V)$ locally at ξ. So, when ξ is a locally complete intersection n-uplet, ξ is defined by the ideal $I(\xi) = (f_1, f_2, f_3)$ of $\mathcal{O}_V$. For $j = 1, 2, 3$, let $F_j = f_j + \sum_{i=1}^{n} a_{ji} e_i$, where the parameters a_{ji} are elements of $\mathbb{C}$. Then the ideal (F_1, F_2, F_3) defines a n-uplet in the neighborhood of ξ in $H^n(V)$ and (a_{ji}) constitutes a local chart of $H^n(V)$ at ξ.

B.1 Local chart of $H^4(V)$ at an elongated quadruplet

Let q_o be an elongated quadruplet of support a point p of V. In an appropriate local coordinate system (x, y, z) centered at p, the quadruplet q_o is defined by the ideal $I(q_o) = (x^3, xy, y^2, z)$ of $\mathcal{O}_V$ (cf. definition 7.(iii), p. 67). Let $I(q_o) = (f_1, f_2, f_3, f_4)$. In this case, the quotient $\dfrac{\mathcal{O}_V}{I(q_o)}$ is a $\mathbb{C}$-vector space of basis $\{\overline{1}, \overline{x}, \overline{x^2}, \overline{y}\}$ over $\mathbb{C}$. Let

$$
\begin{aligned}
F_1 &= x^3 + a_1 x^2 + a_2 x + a_3 y + a_4 \\
F_2 &= xy + a_5 x^2 + a_6 x + a_7 y + a_8 \\
F_3 &= y^2 + a_9 x^2 + a_{10} x + a_{11} y + a_{12} \\
F_4 &= z + a_{13} x^2 + a_{14} x + a_{15} y + a_{16}
\end{aligned}
$$

The ideal $I(q_o)$ possesses five syzygies. The coefficients of these syzygies are given by the columns of the matrix E :

$$
E = \begin{bmatrix}
0 & 0 & 0 & -y & -z \\
-y & 0 & -z & x^2 & 0 \\
x & -z & 0 & 0 & 0 \\
0 & y^2 & xy & 0 & x^3
\end{bmatrix}
$$

Then, each column E_j represents a relation between the f_i. Using **Macaulay** and following the previous method, one establishes the equations in $\mathbb{C}[a_1, \ldots, a_{16}]$, which define $H^4(V)$ locally at q_0. The equations are :

$$
\begin{aligned}
a_4 &= 2a_3a_5a_7 - a_1a_7{}^2 + a_7{}^3 - a_3a_6 + a_2a_7 + a_3a_{11} \\
a_8 &= -a_3a_5{}^2 - a_5a_7{}^2 + a_6a_7 - a_3a_9 \\
a_{10} &= a_1a_5{}^2 + a_5{}^2a_7 - 2a_5a_6 + a_1a_9 - a_7a_9 + a_5a_{11} \\
a_{12} &= a_3a_5{}^3 - a_1a_5{}^2a_7 + a_2a_5{}^2 + 2a_5a_6a_7 + a_3a_5a_9 - a_1a_7a_9 + \\
& \quad a_7{}^2a_9 - a_5a_7a_{11} - a_6{}^2 + a_2a_9 + a_6a_{11}
\end{aligned}
$$

The parameters a_4, a_8, a_{10} and a_{12} are explicit functions of the parameters $a_1, a_2, a_3, a_5,$ a_6, a_7, a_9 and a_{11}. Therefore, $(\underline{a}) = (a_1, a_2, a_3, a_5, a_6, a_7, a_9, a_{11}, a_{13}, a_{14}, a_{15}, a_{16})$ constitutes a chart of $H^4(V)$ at q_0.

B.2 Local equations of $H^4(V)$ at a spherical quadruplet

If now q_0 denotes the spherical quadruplet of support the point p of V, q_0 is defined by the ideal $(x^2, xy, xz, y^2, yz, z^2)$ where (x, y, z) is a local system of coordinates centered at p. Let $q_0 = (f_1, \ldots, f_6)$. The quotient $\dfrac{\mathcal{O}_V}{I(q_0)}$ is in this case a $\mathbb{C}$-vector space of basis $\{\overline{1}, \overline{x}, \overline{y}, \overline{z}\}$. One considers a deformation $F_1, \ldots, F_6$ of the generators $f_1, \ldots, f_6$ in the cobasis :

$$
\begin{aligned}
F_1 &= x^2 + a_1x + a_2y + a_3z + a_4 \\
F_2 &= xy + a_5x + a_6y + a_7z + a_8 \\
F_3 &= xz + a_9x + a_{10}y + a_{11}z + a_{12} \\
F_4 &= y^2 + a_{13}x + a_{14}y + a_{15}z + a_{16} \\
F_5 &= yz + a_{17}x + a_{18}y + a_{19}z + a_{20} \\
F_6 &= z^2 + a_{21}x + a_{22}y + a_{23}z + a_{24}
\end{aligned}
$$

where the parameters a_i are elements of $\mathbb{C}$. The ideal $I(q_0)$ has eight linearly independent syzygies. The coefficients of these syzygies are given by the columns of the matrix E :

$$
E = \begin{bmatrix}
0 & 0 & 0 & 0 & 0 & -z & 0 & -y \\
0 & 0 & 0 & 0 & -y & 0 & -z & x \\
-z & 0 & -y & 0 & 0 & x & y & 0 \\
0 & 0 & 0 & -z & x & 0 & 0 & 0 \\
0 & -z & x & y & 0 & 0 & 0 & 0 \\
x & y & 0 & 0 & 0 & 0 & 0 & 0
\end{bmatrix}
$$

Using **Macaulay** again, one establishes the following equations of the flattener :

$$
-a_4 - a_3a_9 - a_7a_{10} + a_1a_{11} - a_{11}{}^2 + a_2a_{19} + a_3a_{23} = 0
$$

$$-a_8 + a_5 a_6 - a_2 a_{13} - a_{10} a_{15} + a_7 a_{18} = 0$$

$$-a_{12} + a_6 a_9 - a_5 a_{10} + a_{10} a_{14} - a_6 a_{18} + a_{11} a_{18} - a_7 a_{22} = 0$$

$$-a_{16} + a_{11} a_{13} - a_7 a_{17} - a_{15} a_{18} + a_{14} a_{19} - a_{19}^2 + a_{15} a_{23} = 0$$

$$-a_{20} + a_{11} a_{17} + a_{18} a_{19} - a_7 a_{21} - a_{15} a_{22} = 0$$

$$-a_{24} - a_9^2 - a_{10} a_{17} + a_1 a_{21} - a_{11} a_{21} + a_5 a_{22} + a_9 a_{23} = 0$$

$$a_{10} a_{13} - a_6 a_{17} + a_{11} a_{17} - a_7 a_{21} = 0$$

$$a_9 a_{17} + a_{17} a_{18} - a_5 a_{21} + a_{19} a_{21} - a_{13} a_{22} - a_{17} a_{23} = 0$$

$$a_9 a_{13} - a_5 a_{17} + a_{14} a_{17} - a_{13} a_{18} - a_{17} a_{19} + a_{15} a_{21} = 0$$

$$a_9 a_{10} + a_{10} a_{18} - a_2 a_{21} - a_6 a_{22} + a_{11} a_{22} - a_{10} a_{23} = 0$$

$$a_9^2 - a_{18}^2 - a_1 a_{21} + a_6 a_{21} + a_{11} a_{21} - a_5 a_{22} + a_{14} a_{22}$$
$$-a_{19} a_{22} - a_9 a_{23} + a_{18} a_{23} = 0$$

$$a_7 a_9 + a_{10} a_{15} - a_3 a_{17} - a_7 a_{18} = 0$$

$$a_6 a_9 - a_9 a_{11} + a_{10} a_{14} - a_2 a_{17} - a_6 a_{18} + a_{11} a_{18} - 2a_{10} a_{19}$$
$$+a_3 a_{21} + a_7 a_{21} = 0$$

$$a_5 a_{10} - a_2 a_{17} - a_{10} a_{19} + a_7 a_{22} = 0$$

$$a_5 a_9 - a_1 a_{17} + a_6 a_{17} + a_{11} a_{17} - a_5 a_{18} - a_9 a_{19} + a_{18} a_{19}$$
$$-a_{15} a_{22} = 0$$

$$a_5 a_7 - a_3 a_{13} - a_7 a_{14} + a_6 a_{15} - a_{11} a_{15} + a_7 a_{19} = 0$$

$$a_5 a_6 - a_5 a_{11} - a_2 a_{13} - a_{10} a_{15} + a_3 a_{17} + 2a_7 a_{18} - a_6 a_{19}$$
$$+a_{11} a_{19} - a_7 a_{23} = 0$$

$$a_5^2 - a_1 a_{13} + a_6 a_{13} + a_{11} a_{13} - a_5 a_{14} - a_9 a_{15}$$
$$-a_{15} a_{18} + a_{14} a_{19} - a_{19}^2 + a_{15} a_{23} = 0$$

$$a_3 a_5 - a_1 a_7 + a_6 a_7 + a_7 a_{11} - a_2 a_{15} + a_3 a_{19} = 0$$

$$a_2 a_9 - a_1 a_{10} + a_6 a_{10} + a_{10} a_{11} - a_2 a_{18} - a_3 a_{22} = 0$$

$$a_2 a_5 - a_1 a_6 + a_6^2 - a_3 a_9 + a_1 a_{11} - a_{11}^2 - a_2 a_{14} - a_3 a_{18}$$
$$+a_2 a_{19} + a_3 a_{23} = 0$$

Let $(\underline{a}) = (a_1, \ldots, \check{a}_4, \ldots, \check{a}_8, \ldots, \check{a}_{12}, \ldots, \check{a}_{16}, \ldots, \check{a}_{20}, \ldots, \check{a}_{24})$ (the notation $\check{a}_i$ means that a_i is omitted). From the first six equations, the parameters $a_4, a_8, a_{12}, a_{16}, a_{20}$ and a_{24} are explicit functions of the other eighteen parameters $\underline{a}$. These eighteen other parameters $\underline{a}$ are linked by the last fifteen quadratic equations. One checks that these fifteen equations are linearly independent. Therefore, these fifteen quadratic equations of $\mathbb{C}[\underline{a}]$ constitute the local equations of $H^4(V)$ at q_o. (Recall ([I2], [F]) that the spherical quadruplets are the singular points of the Hilbert scheme $H^4(V)$.)

Bibliography

[B1]   BRIANÇON, J. *Description de $Hilb^n\mathbb{C}\{x, y\}$*. Invent. Math. **41** (1977), 45–89.

[B2]   BRIANÇON, J. *Weierstrass préparé à la Hironaka.* Astérisque **7** et **8** (1973), 67–76.

[BGS]   BRIANÇON, J., GRANGER, M., & SPEDER, J. P. *Sur le schéma de Hilbert d'une courbe plane.* Ann. Sci. Ecole Norm. Sup. **14** (1981).

[Co]   COLLEY, S. *Enumerating stationary multiple-points.* Adv. Math. **66** (1987), 149–170.

[E]   EISENBUD, D. "Commutative algebra with a view toward algebraic geometry", Graduate texts in Mathematics **150**, Springer–Verlag (1995).

[ELB]   ELENCWAJG, G. & LE BARZ, P. *Explicit computations in $Hilb^3(\mathbb{P}^2)$*, in "Algebraic Geometry – Sundance 1986," Holme, A. , Speiser, R. Eds. Lecture Notes in Math. **1311**, Springer, 1988, 76–100.

[F]   FOGARTY, J. *Algebraic families on an algebraic surface.* Amer. J. Math. **90** (1967), 511–521.

[FU1]   FULTON, W. Personal letter to P. LE BARZ (01/06/1987).

[FU2]   FULTON, W. "Intersection Theory," Ergebnisse der Mathematik, Springer–Verlag (1984).

[FL]   FULTON, W. & LAKSOV, D. *Residual intersections and the double point formula,* in "Real and Complex Singularities," Proc. Conf., Oslo 1976, Holm, P., Ed., Sijthoff & Noordhoff, 1977, 171–178.

[FMP]   FULTON, W. & MAC PHERSON, R. *A compactification of configuration spaces,* Ann. Math. **139** (1994), 183–225.

[Ga]   GAFFNEY, T. *Punctual Hilbert schemes and resolution of multiple-point singularities.*

[Gal]   GALLIGO, A. *Théorème de division et stabilité en géométrie analytique lo-cale.* Ann. Inst. Fourier, Grenoble, **29**, 2 (1979), 107–184.

[Gr]   GRANGER, M. *Géométrie des schémas de Hilbert ponctuels*, Mém. Soc. Math. France **712** (1982-83).

[G]   GROTHENDIECK, A. *Les schémas de Hilbert*, Séminaire Bourbaki n°221, IHP Paris (1961).

[H]   HARTSHORNE, R. "Algebraic Geometry," Graduate texts in Mathematics, Springer–Verlag (1977).

[I1]   IARROBINO, A. *Hilbert scheme of points: overview of last ten years*, in "Algebraic Geometry – Bowdoin 1985. Part 2," Proc. Sympos. Pure Math. **46** (1987), 297–320.

[I2]   IARROBINO, A. *Reducibility of the Families of 0-Dimensional Schemes on a Variety*, Invent. Math. **15** (1972), 72–77.

[Ka1]   KATZ, S. *Iteration of multiple point formulas and applications to conics*, in "Algebraic Geometry – Sundance 1986," Holme, A., Speiser, R. Eds. Lecture Notes in Math. **1311**, Springer–Verlag, 1988, 147–155.

[Ka2]   KATZ, S. *The desingularization of $Hilb^4(P^3)$ and its Betti numbers*, in "Zero-dimensional schemes," (Proc. Conf., Ravello, 1992).

[Ke]   KEEL, S. *Functorial construction of Le Barz's triangle space with Applications*, Trans. Am. Soc. **335** (1993), 213–229.

[KL1]   KLEIMAN, S. *Multiple-point formulas I: Iteration*, Acta Math. **147** (1981), 13–49.

[KL2]   KLEIMAN, S. *Multiple-point formulas for maps*, in "Enumerative and classical Algebraic Geometry," (Proc. Conf., Nice, 1981), Le Barz, P., Hervier, Y., Eds., Progr. Math. **24**, Birkhäuser, 1982, 237–252.

[KL3]   KLEIMAN, S. *Multiple-point formulas II: the Hilbert scheme*, Sitges (1987), Lecture Notes in Math. **1436** (1990), 101–138.

[KL4]   KLEIMAN, S., with the collaboration of Thorup A. on §3, *Intersection Theory and Enumerative Geometry: A Decade in Review*, in "Algebraic Geometry – Bowdoin 1985. Part 2," Proc. Sympos. Pure Math. **46** (1987), 321–370.

[La]   LAKSOV, D. *Residual intersections and Todd's formula for double locus of a morphism*, Acta Math. **140** (1978), 75–92.

[LB1]   LE BARZ, P. *La variété des triplets complets*, Duke Math. J. **57** (1988), 925–946.

[LB2]   LE BARZ, P. *Contribution des droites d'une surface à ses multisécantes*, Bull. Soc. Math. France **112** (1984), 303–324.

[LB3]   LE BARZ, P. *Un lemme sur les fibrés normaux*, C. R. Acad. Sci. Paris **296** (1983), 911–914.

[Ra]   RAN, Z. *Curvilinear enumerative geometry*, Acta Math. **155** (1985), 81–101.

[Rob]   ROBERTS, J. *Old and new results about the triangle variety*, in "Algebraic Geometry – Sundance 1986," Holme, A., Speiser, R. Eds. Lecture Notes in Math. **1311**, Springer–Verlag, 1988, 197–219.

[Ro]   RONGA, F. *Desingularisation of the triple points and of the stationary points of a map*, Comp. Math. **53** (1984), 211–223.

[RS1]   ROBERTS, J. & SPEISER, R. *Enumerative geometry of triangles I*, Comm. Algebra **12** (1984), 1213–1255.

[RS2]   ROBERTS, J. & SPEISER, R. *Enumerative geometry of triangles II*, Comm. Algebra **14** (1986), 155–191.

[RS3]   ROBERTS, J. & SPEISER, R. *Enumerative geometry of triangles III*, Comm. Algebra **15** (1987), 1929–1966.

[S]   SEMPLE, J. G. *The triangle as a geometric variable*, Mathematika **1** (1954), 80–88.

Index

Index of notation

Printing: Weihert-Druck GmbH, Darmstadt
Binding: Theo Gansert Buchbinderei GmbH, Weinheim